A2-Level Physics

A2 Physics is seriously tricky — no question about that.
To do well, you're going to need to revise properly and practise hard.

This book has thorough notes on all the theory you need,
and it's got practice questions... lots of them.
For every topic there are warm-up and exam-style questions.

And of course, we've done our best to make the whole thing vaguely entertaining for you.

Complete Revision and Practice

Exam Board: AQA A

Editors:
Amy Boutal, Sarah Hilton, Alan Rix, Julie Wakeling, Sarah Williams

Contributors
Peter Cecil, Peter Clarke, Mark A. Edwards, D. Kamya, Barbara Mascetti, John Myers, Zoe Nye, Moira Steven, Andy Williams, Tony Winzor

Proofreaders:
Ian Francis, Glenn Rogers

Published by CGP

ISBN: 978 1 84762 260 0

Groovy website: www.cgpbooks.co.uk
Jolly bits of clipart from CorelDRAW®
Printed by Elanders Ltd, Newcastle upon Tyne.

Based on the classic CGP style created by Richard Parsons.

Contents

How Science Works 2

Unit 4: Section 1 — Further Mechanics

Momentum and Impulse 4
Circular Motion 6
Simple Harmonic Motion 8
Simple Harmonic Oscillators 10
Free and Forced Vibrations 12

Unit 4: Section 2 — Gravitation and Electric Fields

Gravitational Fields 14
Electric Fields 18

Unit 4: Section 3 — Capacitance

Capacitors 22
Charging and Discharging 24

Unit 4: Section 4 — Magnetic Fields

Magnetic Fields 26
Electromagnetic Induction 28
Transformers and Alternators 30

Unit 5: Section 1 — Radioactivity and Nuclear Energy

Scattering to Determine Structure 32
Nuclear Radius and Density 34
Radioactive Emissions 36
Exponential Law of Decay 38
Nuclear Decay 40
Binding Energy 42
Nuclear Fission and Fusion 44

Unit 5: Section 2 — Thermal Physics

Ideal Gases 46
The Pressure of an Ideal Gas 48
Energy and Temperature 50

Unit 5: Option A — Astrophysics

Optical Telescopes 52
Non-Optical Telescopes 55
Distances and Magnitude 58
Stars as Black Bodies 60
Spectral Classes and The H-R Diagram 62
Stellar Evolution 64
The Doppler Effect and Redshift 66
The Big Bang Model of the Universe 68

Unit 5: Option B — Medical Physics

Physics of the Eye 70
Defects of Vision 72
Physics of the Ear 74
Intensity and Loudness 76
Physics of the Heart 78
X-Ray Production 80
X-Ray and MRI Imaging 82
Ultrasound Imaging 84
Endoscopy 86

Unit 5: Option D — Turning Points in Physics

Charge/Mass Ratio of the Electron 88
Millikan's Oil-Drop Experiment 90
Light — Newton vs Huygens 92
Photoelectric Effect 94
Wave-Particle Duality 96
The Speed of Light and Relativity 98
Special Relativity 100

Answering Exam Questions 102

Answers 104

Index 109

Unit 5: Option C — "Applied Physics" isn't covered in this book.

The Scientific Process

'How Science Works' is all about the scientific process — how we develop and test scientific ideas. It's what scientists do all day, every day (well, except at coffee time — never come between a scientist and their coffee).

Scientists Come Up with Theories — Then Test Them...

Science tries to explain **how** and **why** things happen — it **answers questions**. It's all about seeking and gaining **knowledge** about the world around us. Scientists do this by **asking** questions and **suggesting** answers and then **testing** them, to see if they're correct — this is the **scientific process**.

1) **Ask** a question — make an **observation** and ask **why or how** it happens. E.g. what is the nature of light?
2) **Suggest** an answer, or part of an answer, by forming:
   - a **theory** (a possible **explanation** of the observations) e.g. light is a wave.
   - a **model** (a **simplified picture** of what's physically going on)
3) Make a **prediction** or **hypothesis** — a **specific testable statement**, based on the theory, about what will happen in a test situation. E.g. light should interfere and diffract.
4) Carry out a **test** — to provide **evidence** that will support the prediction, or help disprove it. E.g. Young's double-slit experiment.

The evidence supported Quentin's Theory of Flammable Burps.

A theory is only scientific if it can be tested.

...Then They Tell Everyone About Their Results...

The results are **published** — scientists need to let others know about their work. Scientists publish their results in **scientific journals**. These are just like normal magazines, only they contain **scientific reports** (called papers) instead of the latest celebrity gossip.

1) Scientific reports are similar to the **lab write-ups** you do in school. And just as a lab write-up is **reviewed** (marked) by your teacher, reports in scientific journals undergo **peer review** before they're published.
2) The report is sent out to **peers** — other scientists that are experts in the **same area**. They examine the data and results, and if they think that the conclusion is reasonable it's **published**. This makes sure that work published in scientific journals is of a **good standard**.
3) But peer review **can't guarantee** the science is **correct** — other scientists still need to **reproduce** it.
4) Sometimes **mistakes** are made and bad work is published. Peer review **isn't perfect** but it's probably the best way for scientists to self-regulate their work and to publish **quality reports**.

...Then Other Scientists Will Test the Theory Too

Other scientists read the published theories and results, and try to **test the theory** themselves. This involves:

- Repeating the **exact same experiments**.
- Using the theory to make **new predictions** and then testing them with **new experiments**.

If the Evidence Supports a Theory, It's Accepted — for Now

1) If all the experiments in all the world provide evidence to back it up, the theory is thought of as **scientific 'fact'** (for now).
2) But they never become **totally undisputable** fact. Scientific **breakthroughs or advances** could provide new ways to question and test the theory, which could lead to **new evidence** that **conflicts** with the current evidence. Then the testing starts all over again...

And this, my friend, is the **tentative nature of scientific knowledge** — it's always **changing** and **evolving**.

The Scientific Process

So scientists need evidence to back up their theories. They get it by carrying out experiments, and when that's not possible they carry out studies. But why bother with science at all? We want to know as much as possible so we can use it to try and improve our lives (and because we're nosey).

Evidence Comes From **Controlled Lab Experiments...**

1) Results from **controlled experiments** in **laboratories** are **great.**
2) A lab is the easiest place to **control variables** so that they're all **kept constant** (except for the one you're investigating).

For example, finding the charge stored on a capacitor by charging at a constant current and measuring the voltage across it (see p. 22). All other variables need to be kept the same, e.g. the current you use and the temperature, as they may also affect its capacitance.

... That You can Draw **Meaningful Conclusions** From

1) You always need to make your experiments as **controlled** as possible so you can be confident that any effects you see are linked to the variable you're changing.
2) If you do find a relationship, you need to be careful what you conclude. You need to decide whether the effect you're seeing is **caused** by changing a variable, or whether the two are just **correlated.**

"Right Geoff, you can start the experiment now... I've stopped time..."

Society **Makes Decisions** Based on **Scientific Evidence**

1) Lots of scientific work eventually leads to **important discoveries** or breakthroughs that could **benefit humankind**.
2) These results are **used by society** (that's you, me and everyone else) to **make decisions** — about the way we live, what we eat, what we drive, etc.
3) All sections of society use scientific evidence to make decisions, e.g. politicians use it to devise policies and individuals use science to make decisions about their own lives.

Other factors can **influence** decisions about science or the way science is used:

Economic factors

- Society has to consider the **cost** of implementing changes based on scientific conclusions — e.g. the cost of reducing the UK's carbon emissions to limit the human contribution to **global warming**.
- Scientific research is often **expensive**. E.g. in areas such as astronomy, the Government has to **justify** spending money on a new telescope rather than pumping money into, say, the **NHS** or **schools**.

Social factors

- **Decisions** affect **people's lives** — e.g. when looking for a site to build a **nuclear power station**, you need to consider how it would affect the lives of the people in the **surrounding area**.

Environmental factors

- Many scientists suggest that building **wind farms** would be a **cheap** and **environmentally friendly** way to generate electricity in the future. But some people think that because **wind turbines** can **harm wildlife** such as birds and bats, other methods of generating electricity should be used.

So there you have it — how science works...

Hopefully these pages have given you a nice intro to how science works, e.g. what scientists do to provide you with 'facts'. You need to understand this, as you're expected to know how science works yourself — for the exam and for life.

Momentum and Impulse

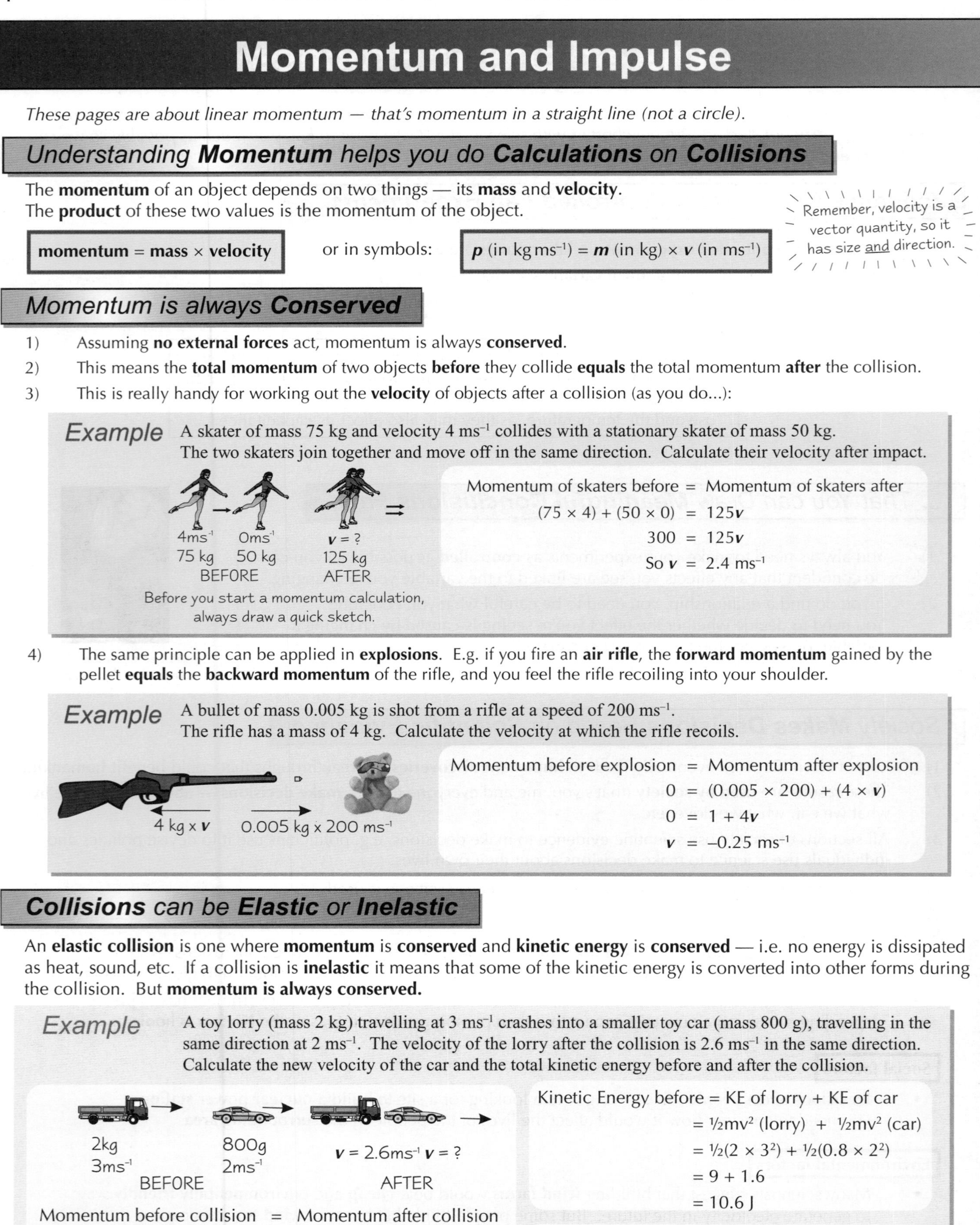

These pages are about linear momentum — that's momentum in a straight line (not a circle).

Understanding **Momentum** helps you do **Calculations** on **Collisions**

The **momentum** of an object depends on two things — its **mass** and **velocity**.
The **product** of these two values is the momentum of the object.

momentum = mass × velocity or in symbols: $\boldsymbol{p}$ (in kg ms^{-1}) = $\boldsymbol{m}$ (in kg) × $\boldsymbol{v}$ (in ms^{-1})

Remember, velocity is a vector quantity, so it has size and direction.

Momentum is always **Conserved**

1) Assuming **no external forces** act, momentum is always **conserved**.
2) This means the **total momentum** of two objects **before** they collide **equals** the total momentum **after** the collision.
3) This is really handy for working out the **velocity** of objects after a collision (as you do...):

Example A skater of mass 75 kg and velocity 4 ms^{-1} collides with a stationary skater of mass 50 kg. The two skaters join together and move off in the same direction. Calculate their velocity after impact.

Before you start a momentum calculation, always draw a quick sketch.

Momentum of skaters before = Momentum of skaters after
$(75 \times 4) + (50 \times 0) = 125\boldsymbol{v}$
$300 = 125\boldsymbol{v}$
So $\boldsymbol{v} = 2.4$ ms^{-1}

4) The same principle can be applied in **explosions**. E.g. if you fire an **air rifle**, the **forward momentum** gained by the pellet **equals** the **backward momentum** of the rifle, and you feel the rifle recoiling into your shoulder.

Example A bullet of mass 0.005 kg is shot from a rifle at a speed of 200 ms^{-1}. The rifle has a mass of 4 kg. Calculate the velocity at which the rifle recoils.

Momentum before explosion = Momentum after explosion
$0 = (0.005 \times 200) + (4 \times \boldsymbol{v})$
$0 = 1 + 4\boldsymbol{v}$
$\boldsymbol{v} = -0.25$ ms^{-1}

Collisions can be **Elastic** or **Inelastic**

An **elastic collision** is one where **momentum** is **conserved** and **kinetic energy** is **conserved** — i.e. no energy is dissipated as heat, sound, etc. If a collision is **inelastic** it means that some of the kinetic energy is converted into other forms during the collision. But **momentum is always conserved.**

Example A toy lorry (mass 2 kg) travelling at 3 ms^{-1} crashes into a smaller toy car (mass 800 g), travelling in the same direction at 2 ms^{-1}. The velocity of the lorry after the collision is 2.6 ms^{-1} in the same direction. Calculate the new velocity of the car and the total kinetic energy before and after the collision.

Momentum before collision = Momentum after collision
$(2 \times 3) + (0.8 \times 2) = (2 \times 2.6) + (0.8\boldsymbol{v})$
$7.6 = 5.2 + 0.8\boldsymbol{v}$
$2.4 = 0.8\boldsymbol{v}$
$\boldsymbol{v} = 3$ ms^{-1}

Kinetic Energy before = KE of lorry + KE of car
$= \frac{1}{2}mv^2$ (lorry) + $\frac{1}{2}mv^2$ (car)
$= \frac{1}{2}(2 \times 3^2) + \frac{1}{2}(0.8 \times 2^2)$
$= 9 + 1.6$
$= 10.6$ J

Kinetic Energy after $= \frac{1}{2}(2 \times 2.6^2) + \frac{1}{2}(0.8 \times 3^2)$
$= 6.76 + 3.6$
$= 10.36$ J

The difference in the two values is the amount of kinetic energy dissipated as heat or sound, or in damaging the vehicles — so this is an inelastic collision.

Momentum and Impulse

Newton's 2nd Law says that Force is the Rate of Change in Momentum...

"The **rate of change of momentum** of an object is **directly proportional** to the **resultant force** which acts on the object." so

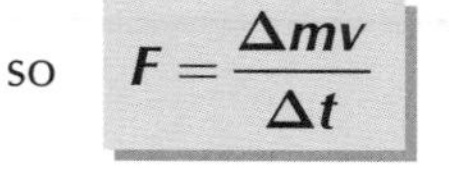

$$F = \frac{\Delta mv}{\Delta t}$$

If mass is constant, this can be written as the well-known equation:

resultant force (F) = mass (m) × acceleration (a)

Learn this — it crops up all over the place in A2 Physics. And learn what it means too:

1) It says that the **more force** you have acting on a certain mass, the **more acceleration** you get.
2) It says that for a given force the **more mass** you have, the **less acceleration** you get.

REMEMBER:
1) The **resultant force** is the **vector sum** of all the forces.
2) The force is **always** measured in **newtons**. **Always**.
3) The **mass** is always measured in **kilograms**.
4) **a** is the **acceleration** of the object as a result of F. It's **always** measured in **metres per second per second** (ms^{-2}).
5) The **acceleration** is always in the **same direction** as the **resultant force**.

Impulse = Change in Momentum

1) Newton's second law says **force = rate of change of momentum**, or $F = (mv - mu) \div t$
2) **Rearranging** Newton's 2nd law gives: $Ft = mv - mu$ (where v is the final velocity and u is the initial velocity), so *impulse = change of momentum*.
   Where **impulse** is defined as **force × time**, Ft.
   The units of impulse are **newton seconds**, Ns.
3) **Impulse** is the **area under** a **force-time graph** — this is really handy for solving problems where the force changes.

Example The graph shows the resultant force acting on a toy car. If the car is initially at rest, what is its momentum after 3 seconds?

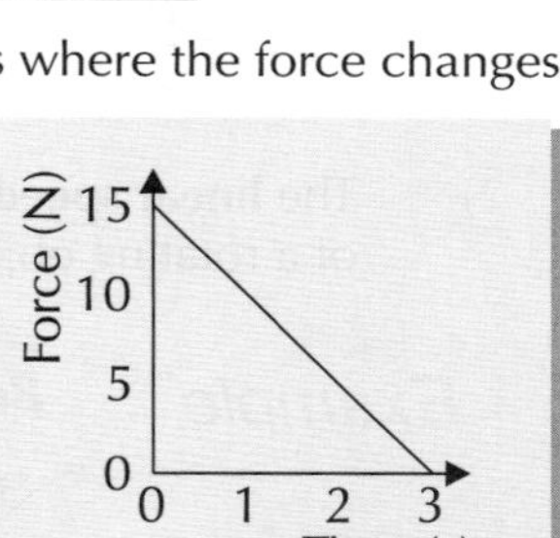

Impulse = **change of momentum** = $mv - mu$. The **initial momentum** (mu) is **zero** because the toy car is stationary to begin with. So, **impulse** = mv.
Impulse is the **area under the graph**, so to find the **momentum** of the car after 3 seconds, you need to find the **area under the graph** between 0 and 3 seconds.
Area under graph = (15 × 3) ÷ 2 = **22.5 Ns**

4) The **force** of an impact can be **reduced** by **increasing the time** of the impact.

Practice Questions

Q1 Give two examples of conservation of momentum in practice.

Q2 Describe what happens when a tiny object makes an elastic collision with a massive object, and why.

Exam Questions

Q1 A ball of mass 0.6 kg moving at 5 ms^{-1} collides with a larger stationary ball of mass 2 kg. The smaller ball rebounds in the opposite direction at 2.4 ms^{-1}.
(a) Calculate the velocity of the larger ball immediately after the collision. [3 marks]
(b) Is this an elastic or inelastic collision? Explain your answer. [3 marks]

Q2 A toy train of mass 0.7 kg, travelling at 0.3 ms^{-1}, collides with a stationary toy carriage of mass 0.4 kg. The two toys couple together. Calculate their new velocity. [3 marks]

Momentum will never be an endangered species — it's always conserved...

It seems a bit of a contradiction to say that momentum's always conserved then tell you that impulse is the change in momentum. The difference is that impulse is only talking about the change of momentum of one of the objects, whereas conservation of momentum is talking about the ***whole*** *system.*

Circular Motion

*It's probably worth putting a bookmark in here — this stuff is needed **all over** the place.*

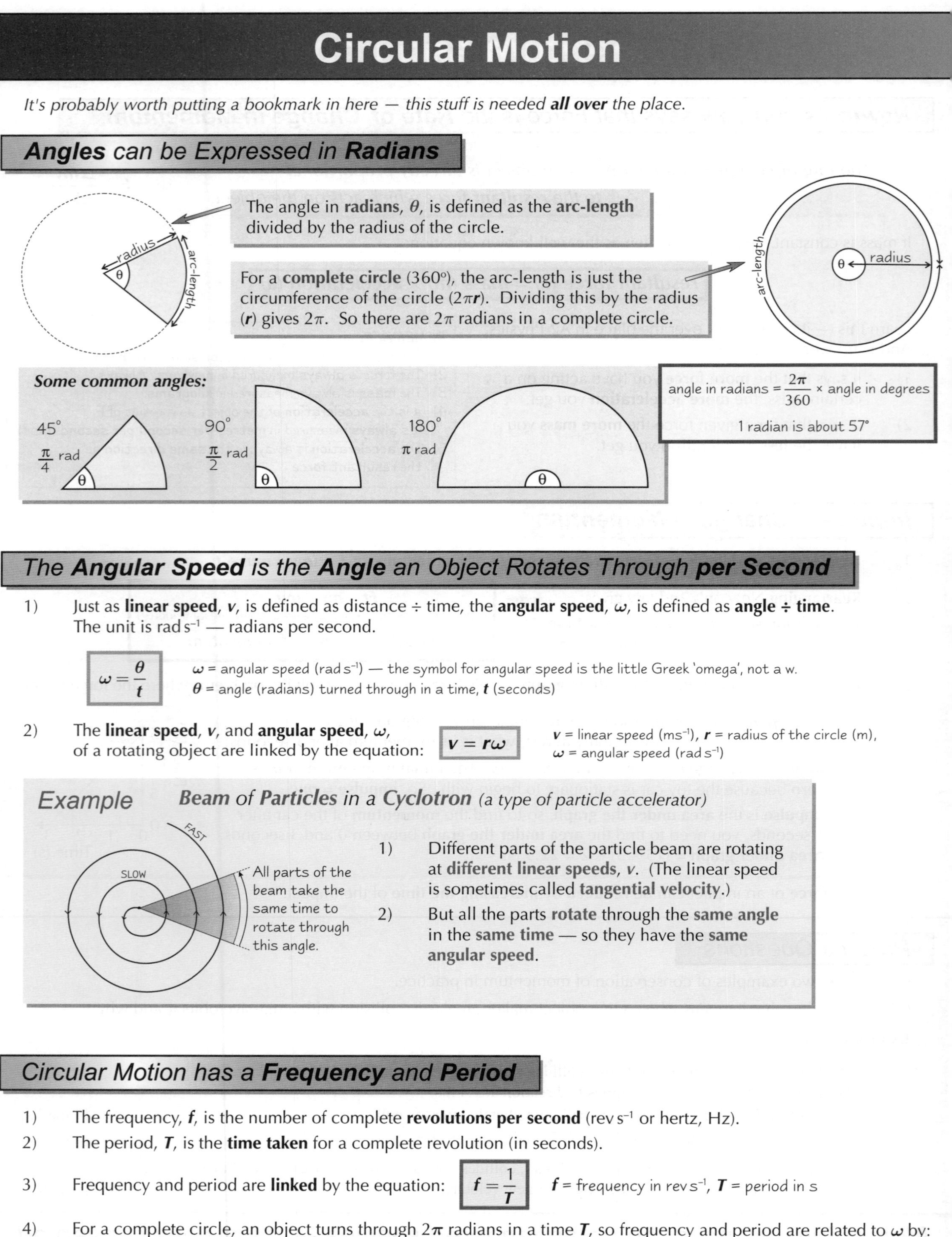

Angles can be Expressed in *Radians*

The angle in **radians**, θ, is defined as the **arc-length** divided by the radius of the circle.

For a **complete circle** (360°), the arc-length is just the circumference of the circle ($2\pi r$). Dividing this by the radius (r) gives 2π. So there are 2π radians in a complete circle.

Some common angles:

45° — $\frac{\pi}{4}$ rad

90° — $\frac{\pi}{2}$ rad

180° — π rad

$$\text{angle in radians} = \frac{2\pi}{360} \times \text{angle in degrees}$$

1 radian is about 57°

The *Angular Speed* is the *Angle* an Object Rotates Through *per Second*

1) Just as **linear speed**, v, is defined as distance ÷ time, the **angular speed**, ω, is defined as **angle ÷ time.** The unit is rad s^{-1} — radians per second.

$$\omega = \frac{\theta}{t}$$

ω = angular speed (rad s^{-1}) — the symbol for angular speed is the little Greek 'omega', not a w.
θ = angle (radians) turned through in a time, t (seconds)

2) The **linear speed**, v, and **angular speed**, ω, of a rotating object are linked by the equation:

$$v = r\omega$$

v = linear speed (ms^{-1}), r = radius of the circle (m), ω = angular speed (rad s^{-1})

Example — *Beam of Particles in a Cyclotron* (a type of particle accelerator)

1) Different parts of the particle beam are rotating at **different linear speeds**, v. (The linear speed is sometimes called **tangential velocity**.)
2) But all the parts **rotate** through the **same angle** in the **same time** — so they have the **same angular speed**.

Circular Motion has a *Frequency* and *Period*

1) The frequency, f, is the number of complete **revolutions per second** (rev s^{-1} or hertz, Hz).
2) The period, T, is the **time taken** for a complete revolution (in seconds).
3) Frequency and period are **linked** by the equation:

$$f = \frac{1}{T}$$

f = frequency in rev s^{-1}, T = period in s

4) For a complete circle, an object turns through 2π radians in a time T, so frequency and period are related to ω by:

$$\omega = 2\pi f \quad \text{and} \quad \omega = \frac{2\pi}{T}$$

f = frequency in rev s^{-1}, T = period in s, ω = angular speed in rad s^{-1}

Circular Motion

Objects Travelling in Circles are **Accelerating** since their **Velocity is Changing**

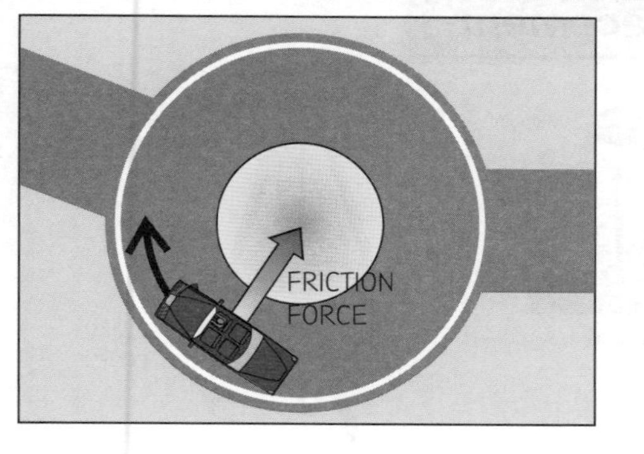

1) Even if the car shown is going at a **constant speed**, its **velocity** is changing since its **direction** is changing.
2) Since acceleration is defined as the **rate of change of velocity**, the car is accelerating even though it isn't going any faster.
3) This acceleration is called the **centripetal acceleration** and is always directed towards the **centre of the circle**.

There are two formulas for centripetal acceleration:

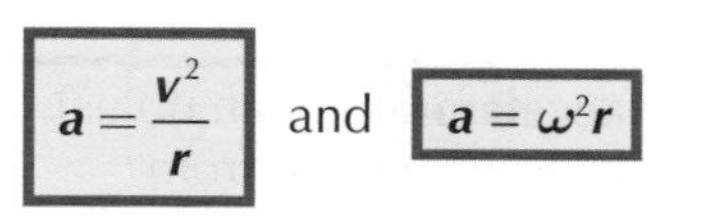

a = centripetal acceleration in ms^{-2}
v = linear speed in ms^{-1}
ω = angular speed in rad s^{-1}
r = radius in m

The **Centripetal Acceleration** is produced by a **Centripetal Force**

From Newton's laws, if there's a **centripetal acceleration**, there must be a **centripetal force** acting towards the **centre of the circle**.

Since ***F*** = ***ma***, the centripetal force must be:

The centripetal force is what keeps the object moving in a circle — remove the force and the object would fly off at a tangent.

Men cowered from the force of the centripede.

Practice Questions

Q1 How many radians are there in a complete circle?
Q2 How is angular speed defined and what is the relationship between angular speed and linear speed?
Q3 Define the period and frequency of circular motion. What is the relationship between period and angular speed?
Q4 In which direction does the centripetal force act, and what happens when this force is removed?

Exam Questions

Q1 (a) At what angular speed does the Earth orbit the Sun? (1 year = 3.2×10^7 s) [2 marks]

(b) Calculate the Earth's linear speed. (Assume radius of orbit = 1.5×10^{11} m) [2 marks]

(c) Calculate the centripetal force needed to keep the Earth in its orbit. (Mass of Earth = 6.0×10^{24} kg) [2 marks]

(d) What is providing this force? [1 mark]

Q2 A bucket full of water, tied to a rope, is being swung around in a vertical circle (so it is upside down at the top of the swing). The radius of the circle is 1 m.

(a) By considering the acceleration due to gravity at the top of the swing, what is the minimum frequency with which the bucket can be swung without any water falling out? [3 marks]

(b) The bucket is now swung with a constant angular speed of 5 rad s^{-1}. What will be the tension in the rope when the bucket is at the top of the swing if the total mass of the bucket and water is 10 kg? [2 marks]

I'm spinnin' around, move out of my way...

*"Centripetal" just means "centre-seeking". The centripetal force is what actually causes circular motion. What you **feel** when you're spinning, though, is the reaction (centrifugal) force. Don't get the two mixed up.*

Simple Harmonic Motion

Something simple at last — I like the sound of this. And colourful graphs too — you're in for a treat here.

SHM is Defined in terms of Acceleration and Displacement

1) An object moving with **simple harmonic motion** (SHM) **oscillates** to and fro, either side of a **midpoint**.
2) The distance of the object from the midpoint is called its **displacement**.
3) There is always a **restoring force** pulling or pushing the object back **towards** the **midpoint**.
4) The **size** of the **restoring force** depends on the **displacement**, and the force makes the object **accelerate** towards the midpoint:

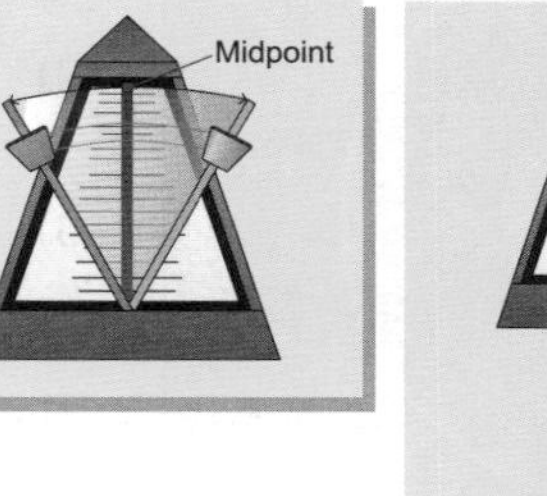

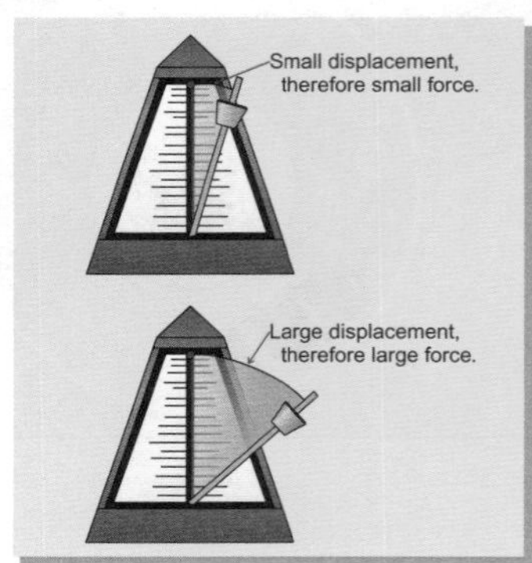

> SHM: an oscillation in which the **acceleration** of an object is **directly proportional** to its **displacement** from the **midpoint**, and is directed **towards the midpoint**.

The Restoring Force makes the Object Exchange PE and KE

1) The **type** of **potential energy** (PE) depends on **what it is** that's providing the **restoring force**. This will be **gravitational PE** for pendulums and **elastic PE** (elastic stored energy) for masses on springs.
2) As the object moves **towards the midpoint**, the restoring force **does work** on the object and so **transfers** some **PE** to **KE**. When the object is moving **away from the midpoint**, all that KE is transferred **back to PE** again.
3) At the **midpoint**, the object's **PE** is **zero** and its **KE** is **maximum**.
4) At the **maximum displacement** (the **amplitude**) on both sides of the midpoint, the object's **KE** is **zero** and its **PE** is **maximum**.

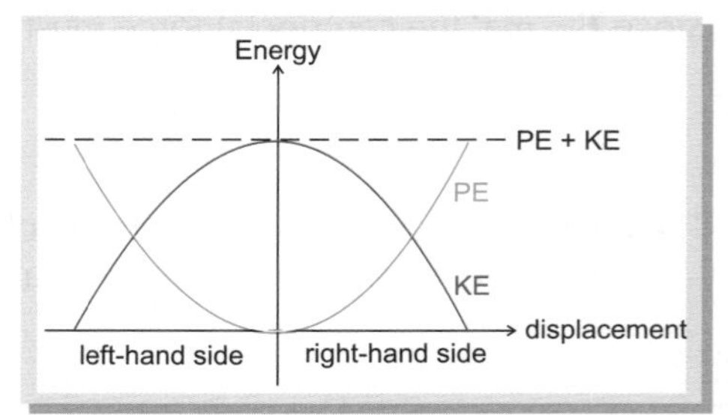

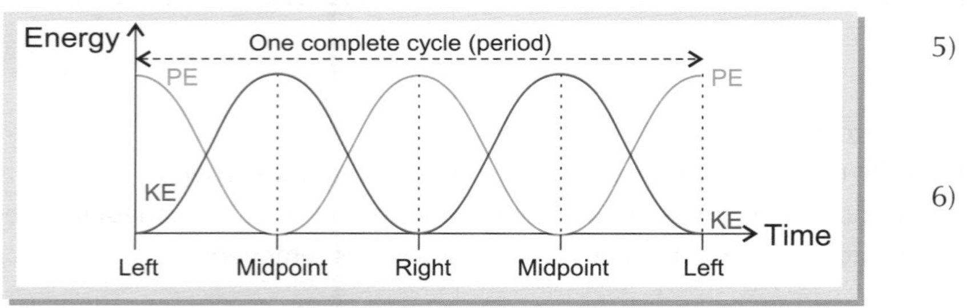

5) The **sum** of the **potential** and **kinetic** energy is called the **mechanical energy** and **stays constant** (as long as the motion isn't damped — see p. 12-13).
6) The **energy transfer** for one complete cycle of oscillation (see graph) is: PE to KE to PE to KE to PE ... and then the process repeats...

You can Draw Graphs to Show Displacement, Velocity and Acceleration

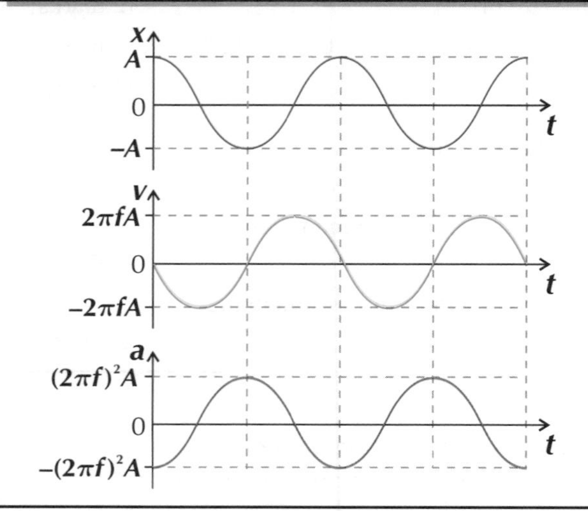

Displacement, x, varies as a cosine or sine wave with a maximum value, A (the amplitude).

Velocity, v, is the gradient of the displacement-time graph. It has a maximum value of $(2\pi f)A$ (where f is the frequency of the oscillation) and is a quarter of a cycle in front of the displacement.

Acceleration, a, is the gradient of the velocity-time graph. It has a maximum value of $(2\pi f)^2A$, and is in antiphase with the displacement.

Simple Harmonic Motion

The Frequency and Period don't depend on the Amplitude

1) From **maximum positive displacement** (e.g. maximum displacement to the right) to **maximum negative displacement** (e.g. maximum displacement to the left) and **back again** is called a **cycle** of oscillation.
2) The **frequency**, f, of the SHM is the number of cycles per second (measured in Hz).
3) The **period**, T, is the **time** taken for a complete cycle (in seconds).

In SHM, the **frequency** and **period** are independent of the **amplitude** (i.e. constant for a given oscillation). So a pendulum clock will keep ticking in regular time intervals even if its swing becomes very small.

Learn the SHM Equations

I know it looks like there are loads of complicated equations to learn here, but don't panic — it's not that bad really. You'll be given these formulas in the exam, so just make sure you know what they mean and how to use them.

1) According to the definition of SHM, the **acceleration**, a, is directly proportional to the **displacement**, x. The **constant of proportionality** depends on the **frequency**, and the acceleration is always in the **opposite direction** from the displacement (so there's a minus sign in the equation).

$$a = -(2\pi f)^2 x$$

Max acceleration: $a_{max} = -(2\pi f)^2 A$

Don't forget, A is the maximum displacement — it's not acceleration.

Jeremy was investigating swinging as a form of simple harmonic motion.

2) The **velocity** is **positive** if the object's moving **away** from the **midpoint**, and **negative** if it's moving **towards** the midpoint — that's why there's a ± sign.

$$v = \pm 2\pi f\sqrt{A^2 - x^2}$$

Max velocity: $v_{max} = 2\pi f A$

3) The **displacement** varies with time according to the equation on the right. To use this equation you need to start timing when the pendulum is at its **maximum displacement** — i.e. when $t = 0$, $x = A$.

$$x = A\cos(2\pi f t)$$

Practice Questions

Q1 Sketch a graph of how the velocity of an object oscillating with SHM varies with time.

Q2 What is the special relationship between the acceleration and the displacement in SHM?

Q3 Given the amplitude and the frequency, how would you work out the maximum acceleration?

Exam Questions

Q1 (a) Define *simple harmonic motion*. [2 marks]

(b) Explain why the motion of a ball bouncing off the ground is not SHM. [1 mark]

Q2 A pendulum is pulled a distance 0.05 m from its midpoint and released.
It oscillates with simple harmonic motion with a frequency of 1.5 Hz. Calculate:

(a) its maximum velocity [1 mark]

(b) its displacement 0.1 s after it is released [2 marks]

(c) the time it takes to fall to 0.01 m from the midpoint after it is released [2 marks]

"Simple" harmonic motion — hmmm, I'm not convinced...

The basic concept of SHM is simple enough (no pun intended). Make sure you can remember the shapes of all the graphs on page 8 and the equations from this page, then just get as much practice at using the equations as you can.

Simple Harmonic Oscillators

There are a couple more equations to learn on this page I'm afraid. The experiment described at the bottom of the page shows where they come from, though, so that should help you remember them.

A **Mass** on a **Spring** is a **Simple Harmonic Oscillator (SHO)**

1) When the mass is **pushed to the left** or **pulled to the right** of the **equilibrium position**, there's a **force** exerted on it.

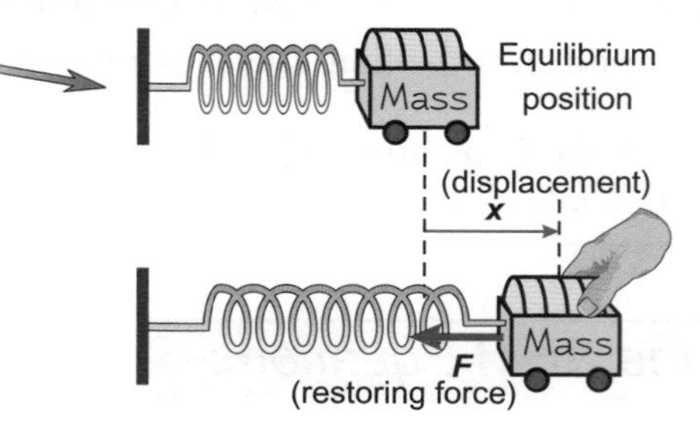

2) The size of this force is:

$$F = -kx$$

where k is the **spring constant** (stiffness) of the spring in Nm^{-1} and x is the displacement in m.

3) After a bit of jiggery-pokery involving Newton's second law and some of the ideas on the previous page, you get the **formula for the period of a mass oscillating on a spring**:

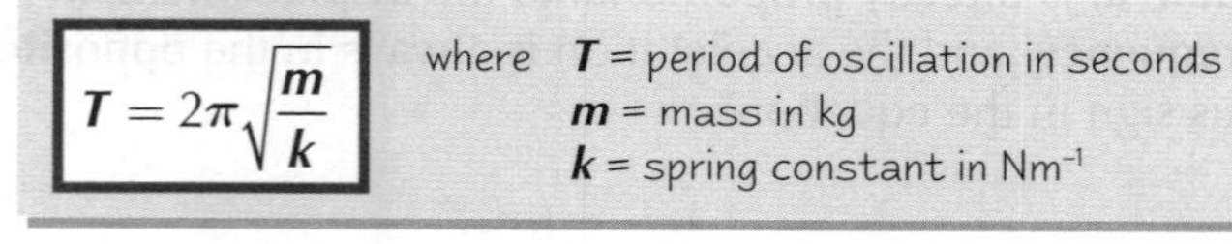

$$T = 2\pi\sqrt{\frac{m}{k}}$$

where T = period of oscillation in seconds
m = mass in kg
k = spring constant in Nm^{-1}

A simple theory of how atoms in a lattice (i.e. a solid) behave can be worked out by considering them as masses oscillating on springs. So there you go.

You Can **Check the Formula Experimentally**

As promised, this experiment shows you where the equations come from.
It's not too tricky — you just have to change **one variable at a time** and see what happens.

Investigating the Mass-Spring System

spring constant k · mass m · trolley · position sensor

1) You could measure the **period**, T, by getting a computer to plot a **displacement-time graph** from a **data logger**.

2) Attach a **trolley** between two **springs**, pull it to one side by a certain amount and then let go. The trolley will **oscillate** back and forth as the springs pull it in each direction.

3) Change the **mass**, m, by loading the trolley with **masses** — don't forget to include the mass of the trolley in your calculations.

4) Change the **spring stiffness**, k, by using different combinations of springs.

$k \rightarrow 2k \rightarrow 3k$; $\frac{1}{2}k$; $\frac{1}{3}k$

5) Change the **amplitude**, A, by pulling the trolley across by different amounts.

6) You'll get the following **results**: *(∝ means "is proportional to")*

a) $T \propto \sqrt{m}$ so $T^2 \propto m$ — graph of T^2 (s^2) against m (kg)

b) $T \propto \sqrt{\frac{1}{k}}$ so $T^2 \propto \frac{1}{k}$ — graph of T^2 (s^2) against $\frac{1}{k}$ (mN^{-1})

c) T doesn't depend on amplitude, A. — graph of T (s) against A (m)

Simple Harmonic Oscillators

The Simple Pendulum is the Classic Example of an SHO

If you set up a simple pendulum attached to an angle sensor and computer like this — then change the length, l, the mass of the bob, m, and the amplitude, A, you get the following results:

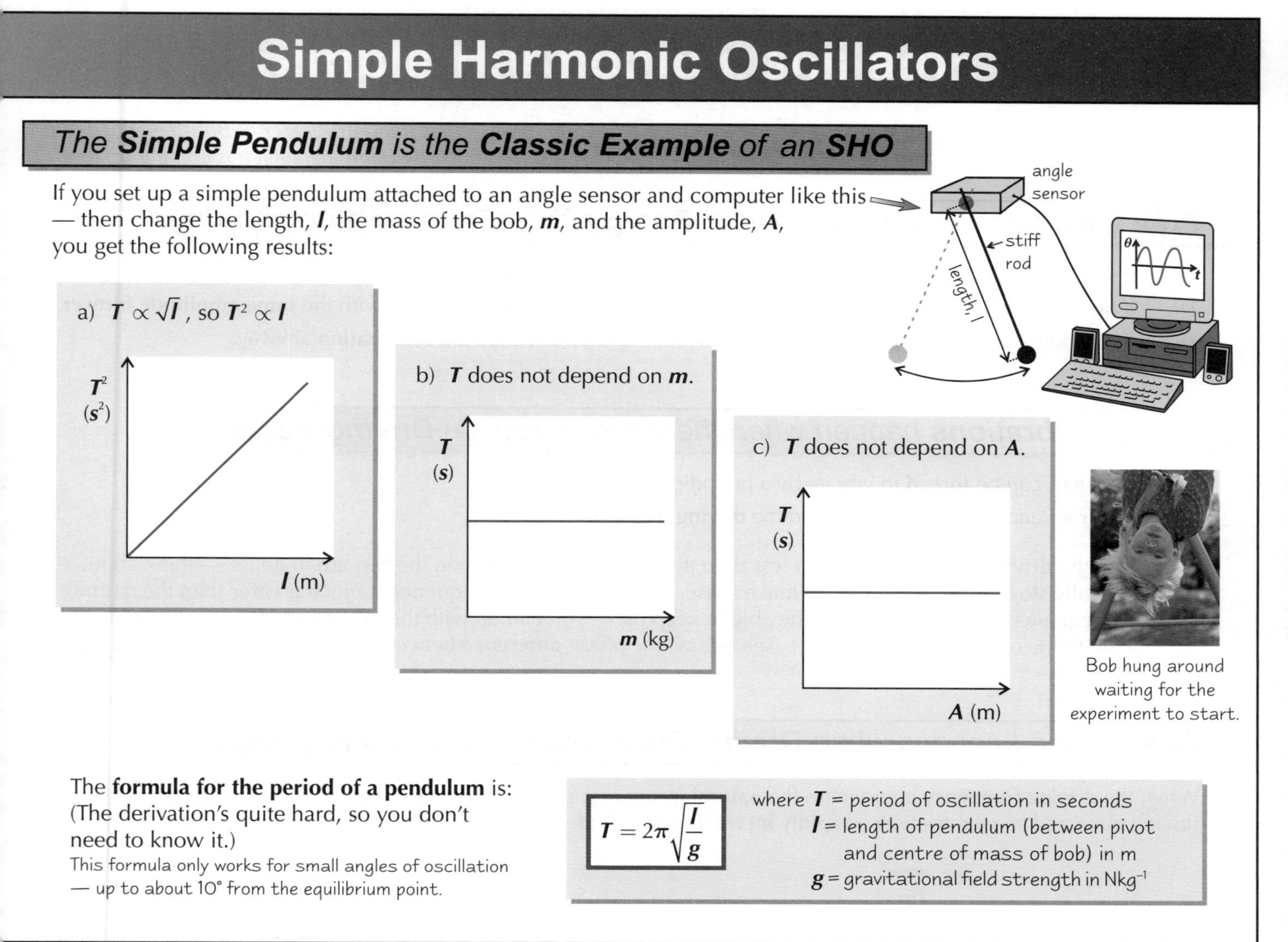

Bob hung around waiting for the experiment to start.

The **formula for the period of a pendulum** is: (The derivation's quite hard, so you don't need to know it.)
This formula only works for small angles of oscillation — up to about 10° from the equilibrium point.

$$T = 2\pi\sqrt{\frac{l}{g}}$$

where T = period of oscillation in seconds
l = length of pendulum (between pivot and centre of mass of bob) in m
g = gravitational field strength in Nkg^{-1}

Practice Questions

Q1 Write down the formulae for the period of a mass on a spring and the period of a pendulum.

Q2 Describe a method you could use to measure the period of an oscillator.

Q3 For a mass-spring system, what graphs could you plot to find out how the period depends on: a) the mass, b) the spring constant, and c) the amplitude? What would they look like?

Exam Questions

Q1 A spring of original length 0.10 m is suspended from a stand and clamp.
A mass of 0.10 kg is attached to the bottom and the spring extends to a total length of 0.20 m.

(a) Calculate the spring constant of the spring in Nm^{-1}. ($g = 9.81\ Nkg^{-1}$) [2 marks]

The spring isn't moving at this point, so the forces on it must be balanced.

(b) The mass is pulled down a further 2 cm and then released. Assuming the mass oscillates with simple harmonic motion, calculate the period of the subsequent oscillations. [1 mark]

(c) What mass would be needed to make the period of oscillation twice as long? [2 marks]

Q2 Two pendulums of different lengths were released from rest at the top of their swing.
It took exactly the same time for the shorter pendulum to make five complete oscillations as it took the longer pendulum to make three complete oscillations.
The shorter pendulum had a length of 0.20 m. Show that the length of the longer one was 0.56 m. [3 marks]

Go on — SHO the examiners what you're made of...

The most important things to remember on these pages are those two period equations. You'll be given them in your exam, but you need to know what they mean and be happy using them.

Free and Forced Vibrations

Resonance... hmm... tricky little beast. Remember the Millennium Bridge, that standard-bearer of British engineering? The wibbles and wobbles were caused by resonance. How was it sorted out? By damping, which is coming up too.

Free Vibrations — No Transfer of Energy To or From the Surroundings

1) If you stretch and release a mass on a spring, it oscillates at its **natural frequency**.
2) If **no energy's transferred** to or from the surroundings, it will **keep** oscillating with the **same amplitude forever**.
3) In practice this **never happens**, but a spring vibrating in air is called a **free vibration** anyway.

Forced Vibrations happen when there's an External Driving Force

1) A system can be **forced** to vibrate by a periodic **external force**.
2) The frequency of this force is called the **driving frequency**.

If the **driving frequency** is much **less than** the **natural frequency** then the two are **in phase** — think about a really slow driver and it should make sense. But, if the **driving frequency** is much **greater than** the **natural frequency**, the oscillator won't be able to keep up — you end up with the driver completely **out of phase** with the oscillator. At **resonance** (see below) the **phase difference** between the driver and oscillator is **90°**.

Resonance happens when Driving Frequency = Natural Frequency

When the **driving frequency** approaches the **natural frequency**, the system gains more and more energy from the driving force and so vibrates with a **rapidly increasing amplitude**. When this happens the system is **resonating**.

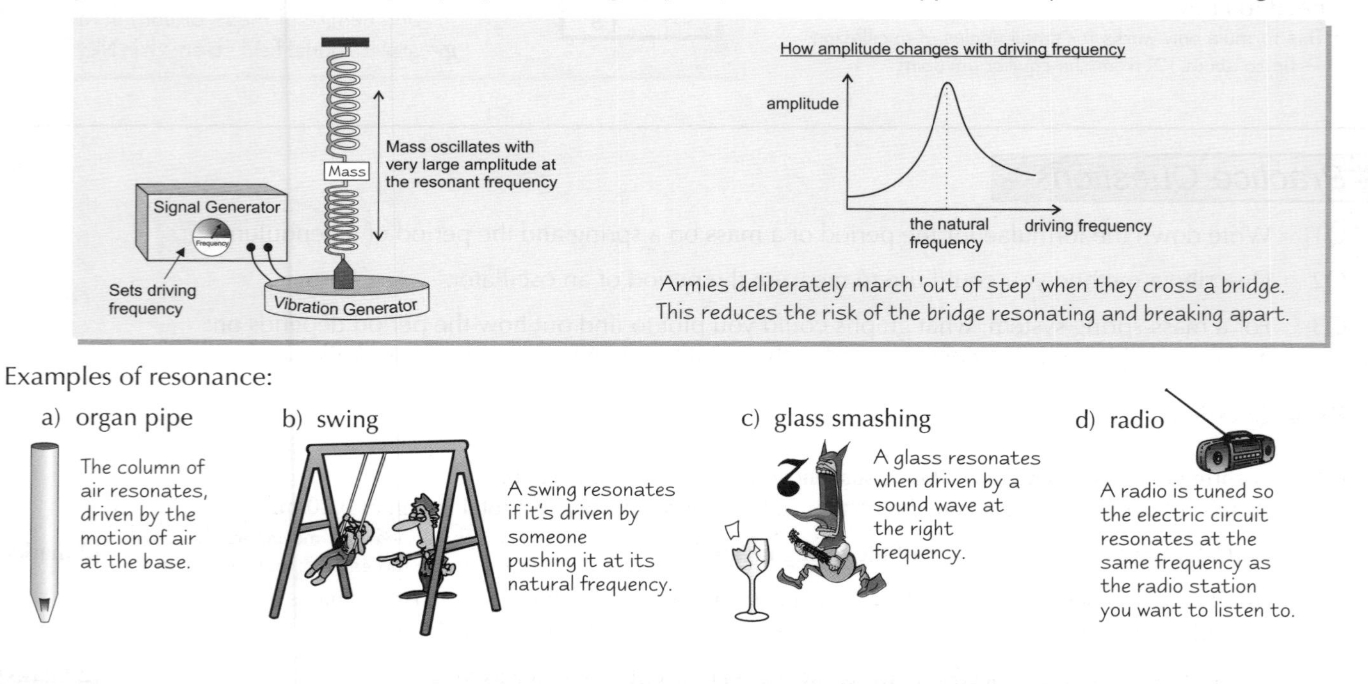

Armies deliberately march 'out of step' when they cross a bridge. This reduces the risk of the bridge resonating and breaking apart.

Damping happens when Energy is Lost to the Surroundings

1) In practice, **any** oscillating system **loses energy** to its surroundings.
2) This is usually down to **frictional forces** like air resistance.
3) These are called **damping forces**.
4) Systems are often **deliberately damped** to **stop** them oscillating or to **minimise** the effect of **resonance**.

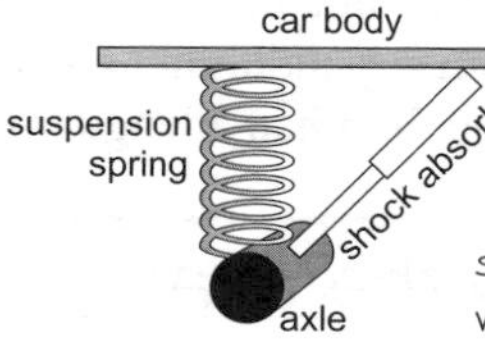

Shock absorbers in a car suspension provide a damping force by squashing oil through a hole when compressed.

Free and Forced Vibrations

Different Amounts of Damping have Different Effects

1) The **degree** of damping can vary from **light** damping (where the damping force is small) to **overdamping**.
2) Damping **reduces** the **amplitude** of the oscillation over time. The **heavier** the damping, the **quicker** the amplitude is reduced to zero.
3) **Critical damping** reduces the amplitude (i.e. stops the system oscillating) in the **shortest possible time**.
4) Car **suspension systems** and moving coil **meters** are critically damped so that they **don't oscillate** but return to equilibrium as quickly as possible.
5) Systems with **even heavier damping** are **overdamped**. They take **longer** to return to equilibrium than a critically damped system.

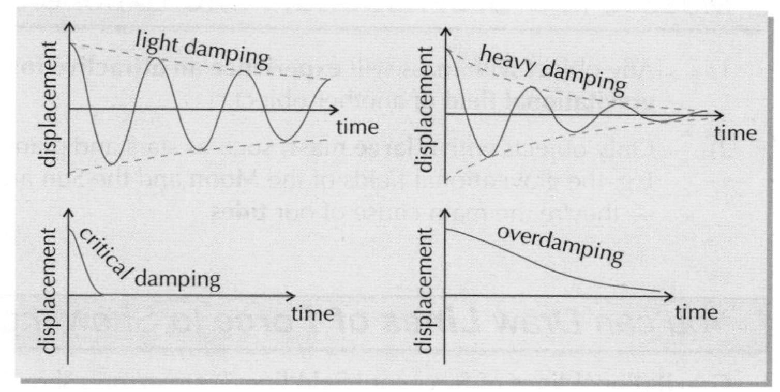

6) **Plastic deformation** of ductile materials **reduces** the **amplitude** of oscillations in the same way as damping. As the material changes shape, it **absorbs energy**, so the oscillation will be smaller.

Damping Affects Resonance too

1) **Lightly damped** systems have a **very sharp** resonance peak. Their amplitude only increases dramatically when the **driving frequency** is **very close** to the **natural frequency**.
2) **Heavily damped** systems have a **flatter response**. Their amplitude doesn't increase very much near the natural frequency and they aren't as **sensitive** to the driving frequency.

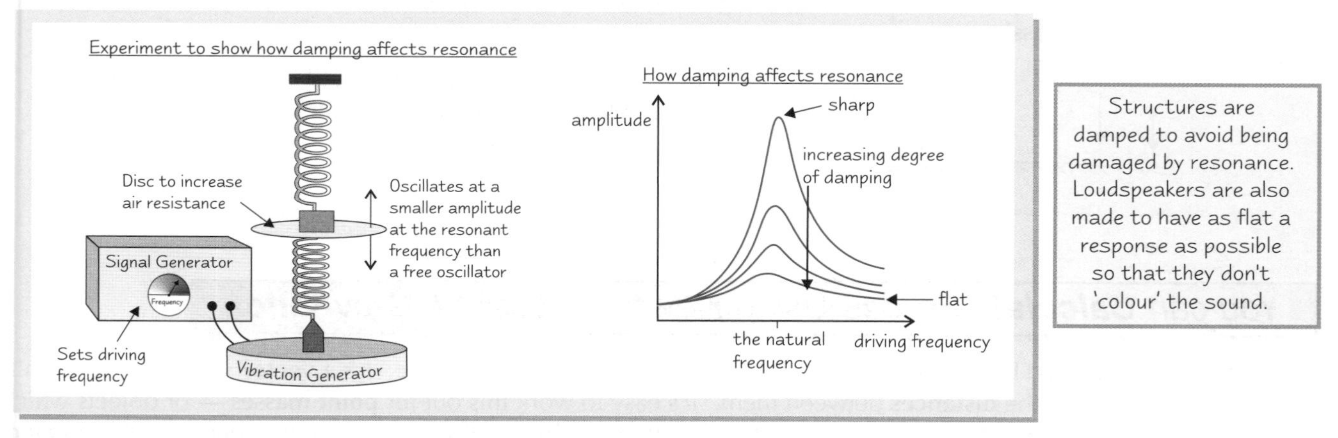

Structures are damped to avoid being damaged by resonance. Loudspeakers are also made to have as flat a response as possible so that they don't 'colour' the sound.

Practice Questions

Q1 What is a free vibration? What is a forced vibration?

Q2 Draw diagrams to show how a damped system oscillates with time when the system is lightly damped and when the system is critically damped.

Exam Questions

Q1 (a) Describe resonance. [2 marks]

(b) Draw a diagram to show how the amplitude of a lightly damped system varies with driving frequency. [2 marks]

(c) On the same diagram, show how the amplitude of the system varies with driving frequency when it is heavily damped. [1 mark]

Q2 (a) Describe critical damping. [1 mark]

(b) State one situation where critical damping is used. [1 mark]

A2 Physics — it can really put a damper on your social life...

Resonance can be really useful (radios, oboes, swings — yay) or very, very bad...

Gravitational Fields

*Gravity's all about masses **attracting** each other. If the Earth didn't have a **gravitational field,** apples wouldn't fall to the ground and you'd probably be floating off into space instead of sitting here reading this page...*

Masses in a Gravitational Field Experience a Force Of Attraction

1) Any object with mass will **experience an attractive force** if you put it in the **gravitational field** of another object.
2) Only objects with a **large** mass, such as stars and planets, have a significant effect. E.g. the gravitational fields of the **Moon** and the **Sun** are noticeable here on Earth — they're the main cause of our **tides**.

Tides are caused by gravitational fields.

You can Draw Lines of Force to Show the Field Around an Object

Gravitational lines of force (or "**field lines**") are **arrows** showing the **direction of the force** that masses would feel in a gravitational field.

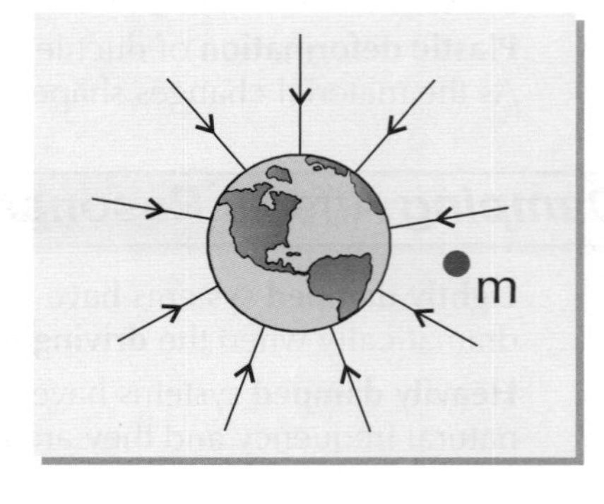

1) If you put a small mass, ***m***, anywhere in the Earth's gravitational field, it will always be attracted **towards** the Earth.
2) The Earth's gravitational field is **radial** — the lines of force meet at the centre of the Earth.
3) If you move mass ***m*** further away from the Earth — where the **lines** of force are **further apart** — the **force** it experiences **decreases**.
4) The small mass, ***m***, has a gravitational field of its own. This doesn't have a noticeable effect on the Earth though, because the Earth is so much **more massive**.

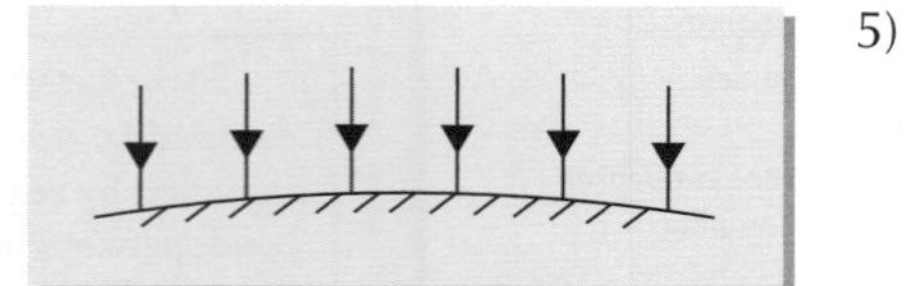

5) Close to the Earth's surface, the field is (almost) uniform — the **field lines** are (almost) **parallel**.

You can Calculate Forces Using Newton's Law of Gravitation

The **force** experienced by an object in a gravitational field is always **attractive**. It's a **vector** which depends on the **masses** involved and the **distances** between them. It's easy to work this out for **point masses** — or objects which behave as if all their mass is concentrated at the centre, e.g. uniform spheres. You just put the numbers into this equation...

NEWTON'S LAW OF GRAVITATION:

$$F = (-)\frac{GMm}{r^2}$$

There's sometimes a negative sign, to show that the vector F is in the opposite direction to r.

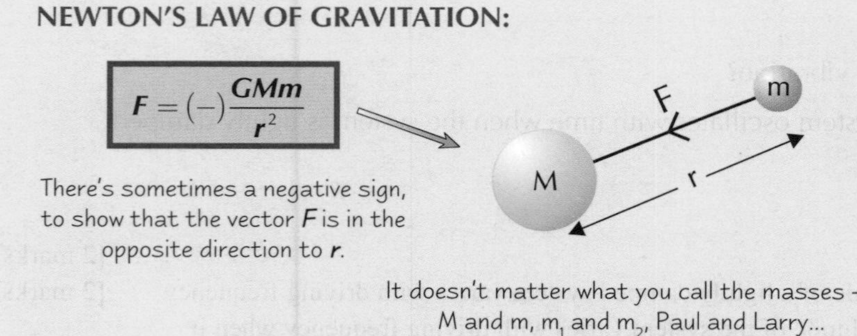

It doesn't matter what you call the masses — M and m, m_1 and m_2, Paul and Larry...

The diagram shows the force acting on m due to M. (The force on M due to m is equal but in the opposite direction.)

M and m behave as point masses.

G is the **gravitational constant** — 6.67×10^{-11} Nm²kg⁻².

r is the distance (in metres) between the centres of the two masses.

The law of gravitation is an **inverse square law** $\left(F \propto \frac{1}{r^2}\right)$ so:

1) if the distance **r** between the masses **increases** then the force **F** will **decrease**.
2) if the **distance doubles** then the **force** will be one **quarter** the strength of the original force.

Gravitational Fields

The Field Strength is the Force per Unit Mass

Gravitational field strength, g, is the **force per unit mass**. Its value depends on **where you are** in the field. There's a really simple equation for working it out:

$$g = \frac{F}{m}$$

g has units of newtons per kilogram (Nkg^{-1})

1) F is the force experienced by a mass m when it's placed in the gravitational field. Divide F by m and you get the **force per unit mass**.
2) g is a **vector** quantity, always pointing towards the centre of the mass whose field you're describing.
3) Since the gravitational field is almost uniform at the Earth's surface, you can assume g is a constant.
4) g is just the **acceleration** of a mass in a gravitational field. It's often called the **acceleration due to gravity**.

The **value** of g at the **Earth's surface** is approximately **9.81** ms^{-2} (or 9.81 Nkg^{-1}).

In a Radial Field, g is Inversely Proportional to r²

Point masses have **radial** gravitational fields (see page 14).
The value of g depends on the distance r from the point mass M...

$$g = \frac{GM}{r^2}$$

where g is the **acceleration due to gravity** (ms^{-2}),
G is the **gravitational constant** (6.67×10^{-11} Nm^2kg^{-2}),
M is a point **mass** and r is the **distance** from the centre (m).

And it's an **inverse square law** again — as r **increases**, g **decreases**.
If you plot a graph of g against r for the **Earth**, you get a curve like this.
It shows that g is greatest at the surface of the Earth, but decreases rapidly as r increases and you move further away from the centre of the Earth.
The **area** under this curve gives you the **gravitational potential**, V — see page 16 for more information.

R_E is the **radius** of the Earth.

g
R_E
r

Practice Questions

Q1 Write down Newton's law of gravitation.

Q2 Sketch a graph of distance from mass (r) against gravitational field strength (g) for a point mass.

Exam Questions

Q1 The Earth's radius is approximately 6400 km. The mass of the Sun is 1.99×10^{30} kg. The average distance from the Earth to the Sun is 1.5×10^{11} m.

(a) Estimate the mass of the Earth (use g = 9.81 Nkg^{-1} at the Earth's surface). [2 marks]

(b) Estimate the force of gravitational attraction between the Sun and the Earth. [2 marks]

Q2 The Moon has a mass of 7.35×10^{22} kg and a radius of 1740 km.

(a) Calculate the value of g at the Moon's surface. [1 mark]

(b) Calculate the force acting on a 25 kg mass on the Moon's surface. [1 mark]

If you're really stuck, put 'Inverse Square Law'...

Clever chap, Newton, but famously tetchy. He got into fights with other physicists, mainly over planetary motion and calculus... the usual playground squabbles. Then he spent the rest of his life trying to turn scrap metal into gold. Weird.

Gravitational Fields

Gravitational Potential is Potential Energy per Unit Mass

The **gravitational potential**, ***V***, at a point is the **gravitational potential energy** that a **unit mass** at that point would have. For example, if a **1 kg** mass has **10 J of potential energy** at a point **Z**, then the **gravitational potential at Z is 10 Jkg⁻¹**.

In a **radial field** (like the Earth's), the equation for gravitational potential is:

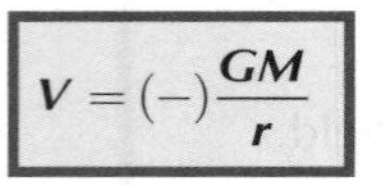

$$V = (-)\frac{GM}{r}$$

where ***V*** is **gravitational potential** (Jkg^{-1}), ***G*** is the **gravitational constant**, ***M*** is the **mass** of the object causing the gravitational field (kg), and ***r*** is the **distance** from the centre of the object (m).

Gravitational potential is **negative** on the **surface** of the mass and **increases with distance** from the mass. This means that the gravitational potential at an **infinite distance** from the mass will be **zero**. The graph shows how **gravitational potential** varies with **distance**.

If you find the **gradient** of this graph at a particular **point**, you get the value of ***g*** at that point.

In other words:

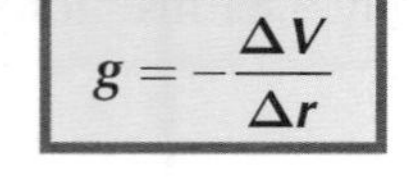

$$g = -\frac{\Delta V}{\Delta r}$$

g is gravitational field strength (Nkg^{-1}) — see page 15.

The gradient of a **tangent** gives the value of ***g*** at that point.

Gravitational Potential Difference is the Energy Needed to Move a Unit Mass

Two points at different distances from a mass will have **different** gravitational potentials (because gravitational potential increases with distance) — this means that there is a **gravitational potential difference** between these two points.

When you **move** an object you do **work** against the force of **gravity** — the **amount of energy** you need depends on the **mass** of the object and the **gravitational potential difference** you move it through:

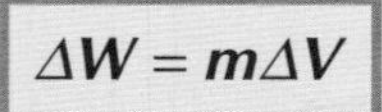

$$\Delta W = m\Delta V$$

where ΔW is the **work done** (J), ***m*** is the **mass** of the object (kg) and ΔV is the **gravitational potential difference** (Jkg^{-1}).

Planets and Satellites are Kept in Orbit by Gravitational Forces

1) A **satellite** is just any **smaller mass** which **orbits** a **much larger mass** — the **Moon** is a satellite of the Earth, **planets** are **satellites** of the **Sun**.
2) **Satellites** are kept in **orbit** by the **gravitational 'pull'** of the mass they're orbiting.
3) In our Solar System, the planets have **nearly circular orbits**, so you can use the **equations of circular motion** (see p.6-7) to investigate their **speed** and **orbital period** (see below).

It's either gravitational pull or a giant white rabbit that makes the Earth orbit the Sun. I know which one I believe...

Orbital Period and Speed Depend on Radius

Any object undergoing **circular motion** (e.g. a **satellite**) is kept in its path by a **centripetal force**. What **causes** this force depends on the object — in the case of **satellites** it's the **gravitational attraction** of the mass they're orbiting. This means that the **centripetal** and **gravitational forces** acting on a satellite must be **equal**:

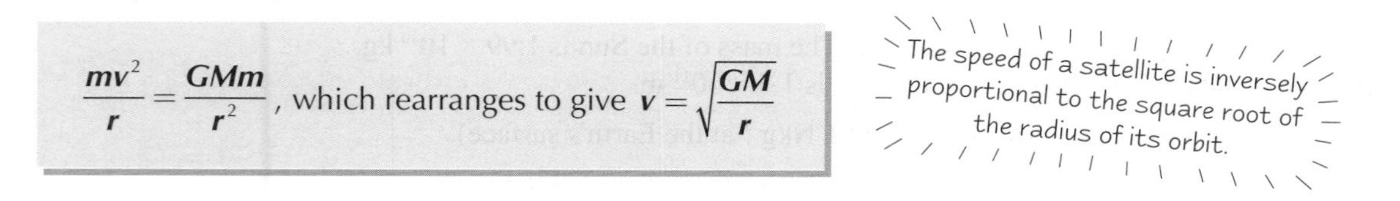

$$\frac{mv^2}{r} = \frac{GMm}{r^2}\text{, which rearranges to give } v = \sqrt{\frac{GM}{r}}$$

The **time** taken for a satellite to make **one orbit** is called the **orbital period**, ***T***. For circular motion, $T = \frac{2\pi r}{v}$.

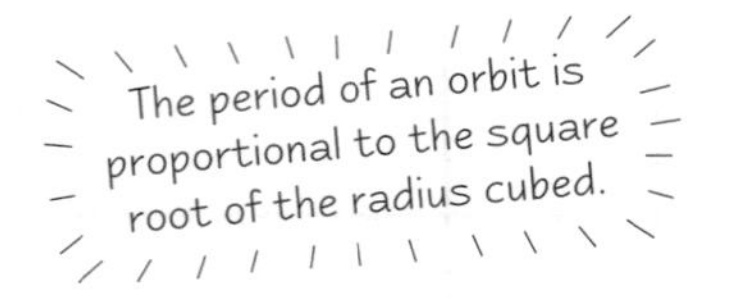

Substitute for ***v*** and rearrange: $T = \sqrt{\frac{4\pi^2 r^3}{GM}}$

The **greater** the **radius** of a satellite's orbit, the **slower** it will travel and the **longer** it will take to complete **one orbit**.

Gravitational Fields

You Can Solve Problems About Orbital Radius, Period and Speed

Example The Moon takes 27.3 days to orbit the Earth. Calculate its distance from the Earth. Take the mass of the Earth to be 5.975 × 10^{24} kg.

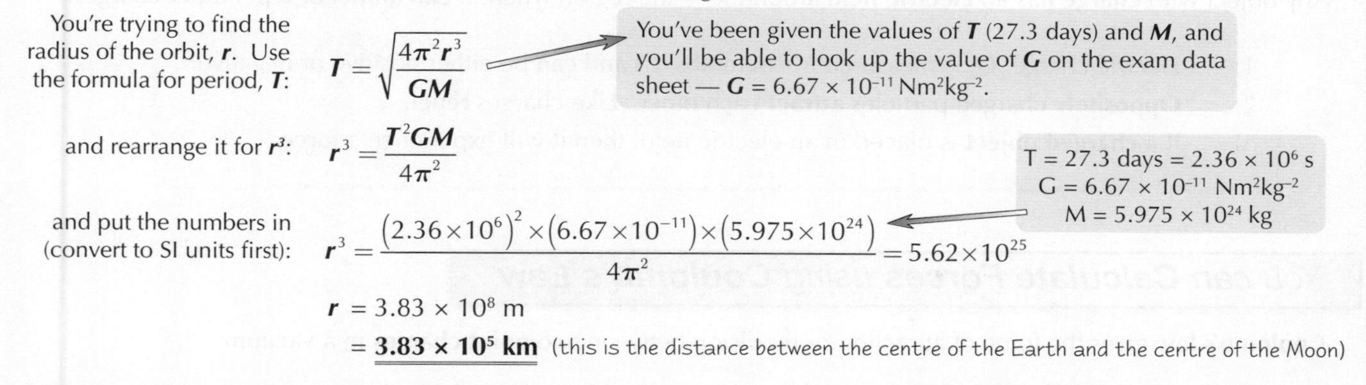

You're trying to find the radius of the orbit, ***r***. Use the formula for period, ***T***: $T = \sqrt{\frac{4\pi^2 r^3}{GM}}$

You've been given the values of ***T*** (27.3 days) and ***M***, and you'll be able to look up the value of ***G*** on the exam data sheet — $G = 6.67 \times 10^{-11}$ Nm²kg⁻².

and rearrange it for r^3: $r^3 = \frac{T^2 GM}{4\pi^2}$

T = 27.3 days = 2.36×10^6 s
G = 6.67×10^{-11} Nm^2kg^{-2}
M = 5.975×10^{24} kg

and put the numbers in (convert to SI units first): $r^3 = \frac{(2.36\times10^6)^2 \times (6.67\times10^{-11}) \times (5.975\times10^{24})}{4\pi^2} = 5.62\times10^{25}$

$r = 3.83 \times 10^8$ m

= **3.83 × 10^5 km** (this is the distance between the centre of the Earth and the centre of the Moon)

The Energy of an Orbiting Satellite is Constant

An orbiting **satellite** has **kinetic** and **potential energy** — its **total energy** (i.e. kinetic + potential) is always **constant**.

1) In a **circular orbit**, a satellite's **speed** and **distance** above the mass it's orbiting are **constant**. This means that its **kinetic energy** and **potential energy** are also both **constant**.
2) In an **elliptical orbit**, a satellite will **speed up** as it's **height decreases** (and slow down as its height increases). This means that its **kinetic energy increases** as its **potential energy decreases** (and vice versa), so the **total energy** remains **constant**.

Geosynchronous Satellites Orbit the Earth once in 24 hours

1) Geosynchronous satellites orbit directly over the **equator** and are **always above the same point** on Earth.
2) A geosynchronous satellite travels at the **same angular speed as the Earth** turns below it.
3) These satellites are really useful for sending TV and telephone signals — the satellite is **stationary** relative to a certain point on the **Earth**, so you don't have to alter the angle of your receiver (or transmitter) to keep up.
4) Their orbit takes exactly **one day**.

Practice Questions

Q1 What is gravitational potential? Write an equation for it.

Q2 The International Space Station orbits the Earth with velocity *v*. If another vehicle docks with it, increasing its mass, what difference, if any, does this make to the speed or radius of the orbit?

Q3 Would a geosynchronous satellite be useful for making observations for weather forecasts? Give reasons.

Exam Questions

(Use G = 6.67 × 10^{-11} Nm²kg⁻², mass of Earth = 5.98 × 10^{24} kg, radius of Earth = 6400 km)

Q1 (a) A satellite orbits 200 km above the Earth's surface. Calculate the period of the satellite's orbit. [2 marks]
(b) Calculate the linear speed of the satellite. [1 mark]

Q2 At what height above the Earth's surface would a geosynchronous satellite orbit? [3 marks]

No fluffy bunnies were harmed in the making of these pages...

When I hear the word 'satellite' I just think of man-made ones, e.g. for mobile phones or TV, and tend to forget that planets (including Earth) and the Moon are satellites too — don't make the same mistake. You're probably best off learning all the stuff about satellites and their orbits too — just knowing what they are won't get you all that far in the exam, I'm afraid.

Electric Fields

Electric fields can be attractive or repulsive, so they're different from gravitational ones. It's all to do with ***charge****.*

There is an Electric Field around a Charged Object

Any object with **charge** has an **electric field** around it — the region where it can attract or repel other charges.

1) Electric charge, Q, is measured in **coulombs** (C) and can be either positive or negative.
2) **Oppositely** charged particles **attract** each other. **Like** charges **repel**.
3) If a **charged object** is placed in an electric field, then it will experience a **force**.

You can Calculate Forces using Coulomb's Law

Coulomb's law gives the force of attraction or repulsion between two **point charges** in a **vacuum**:

COULOMB'S LAW:

$$F = \frac{1}{4\pi\varepsilon_0}\frac{Q_1Q_2}{r^2}$$

ε_0 ("epsilon-nought") is the **permittivity of free space** and is equal to 8.85×10^{-12} Fm^{-1}
Q_1 and Q_2 are the **charges**
r is the **distance** between Q_1 and Q_2

This unit is a 'farad per metre' — see p.22.

1) The force on Q_1 is always **equal** and **opposite** to the force on Q_2 — the **direction** depends on the charges.

If the charges are **opposite** then the force is **attractive**. *F* will be **negative**.

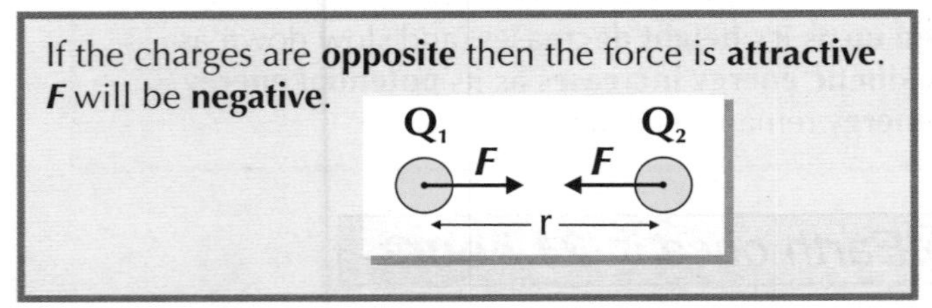

If Q_1 and Q_2 are **alike** then the force is **repulsive**. *F* will be **positive**.

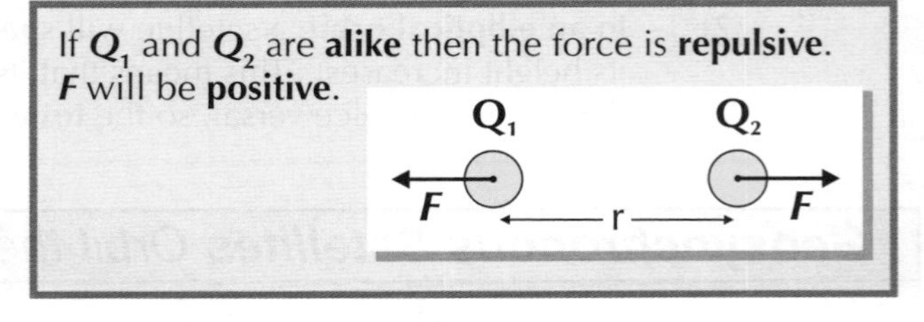

2) Coulomb's law is an **inverse square law**. The further apart the charges, the weaker the force between them.
3) If the point charges aren't in a vacuum, then the size of the force F also depends on the **permittivity**, ε, of the material between them.

Mr Harris liked to explain Coulomb's law using prairie dogs.

Electric Field Strength is Force per Unit Charge

Electric field strength, E, is defined as the **force per unit positive charge**.
It's the force that a charge of +1 C would experience if it was placed in the electric field.

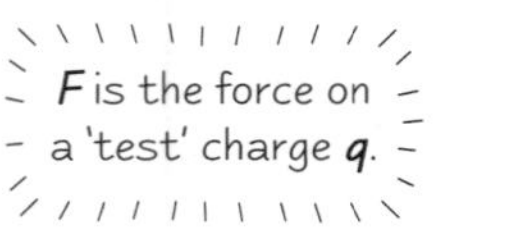

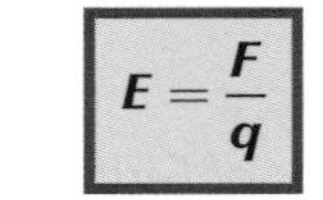

where E is **electric field strength** (NC^{-1}),
F is the **force** (N) and q is the **charge** (C)

1) E is a **vector** pointing in the **direction** that a **positive charge** would **move**.
2) The units of E are **newtons per coulomb** (NC^{-1}).
3) Field strength depends on **where you are** in the field.
4) A **point charge** — or any body that behaves as if all its charge is concentrated at the centre — has a **radial** field.

Electric Fields

In a *Radial Field*, *E* is *Inversely Proportional* to r^2

1) In a **radial field**, the electric field strength, ***E***, depends on the distance ***r*** from the point charge ***Q***:

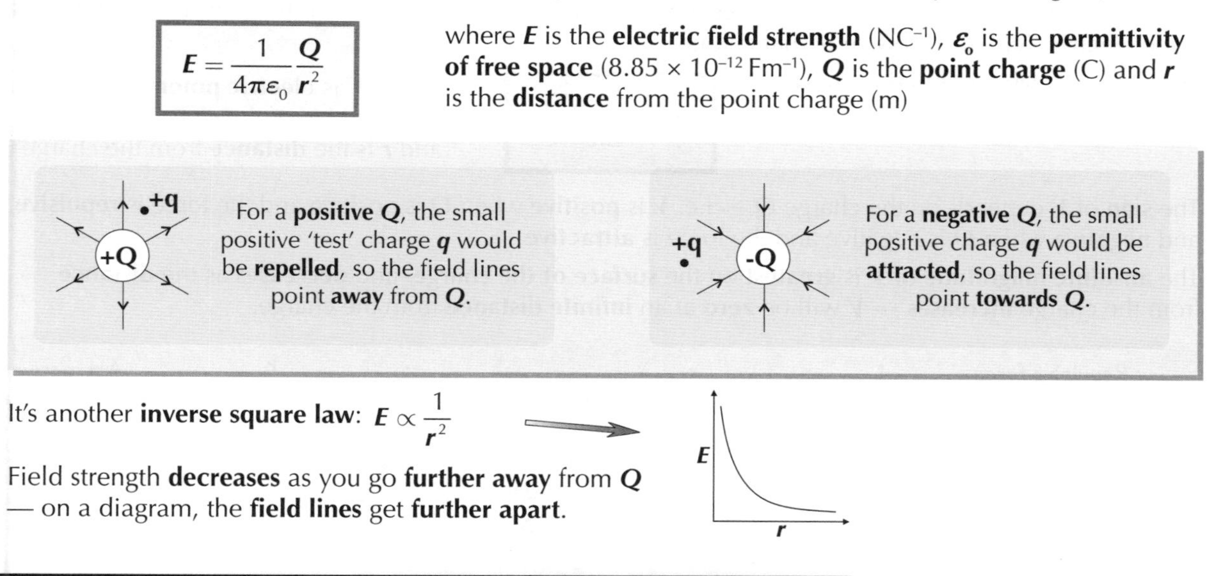

$$E = \frac{1}{4\pi\varepsilon_0}\frac{Q}{r^2}$$

where ***E*** is the **electric field strength** (NC^{-1}), ε_0 is the **permittivity of free space** ($8.85 \times 10^{-12}\ Fm^{-1}$), ***Q*** is the **point charge** (C) and ***r*** is the **distance** from the point charge (m)

2) It's another **inverse square law**: $E \propto \frac{1}{r^2}$

3) Field strength **decreases** as you go **further away** from ***Q*** — on a diagram, the **field lines** get **further apart**.

In a *Uniform Field*, *E* is *Inversely Proportional* to *d*

A **uniform field** can be produced by connecting two **parallel plates** to the opposite poles of a battery.

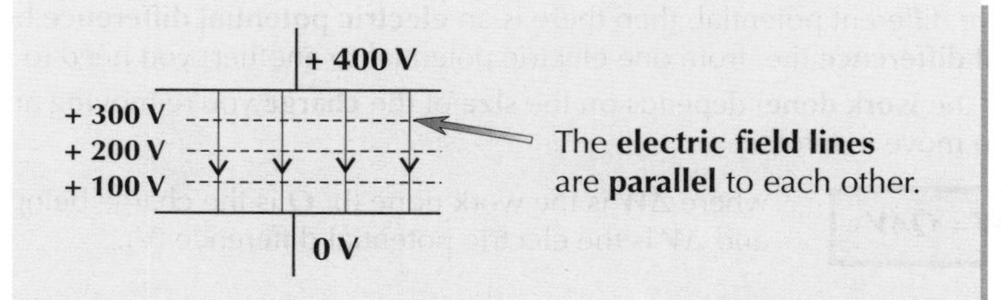

Field strength ***E*** is the **same** at **all points** between the two plates and is given by:

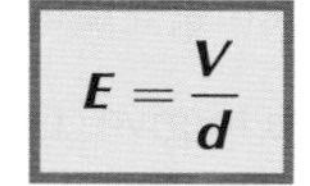

$$E = \frac{V}{d}$$

where ***E*** is the **electric field strength** (Vm^{-1} or NC^{-1}), ***V*** is the **potential difference** between the plates (V) and ***d*** is the distance between them (m)

Practice Questions

Q1 Write down Coulomb's law.

Q2 Sketch a radial electric field and a uniform electric field. How would you find ***E*** for each?

Exam Questions

(Use $\varepsilon_0 = 8.85 \times 10^{-12}\ Fm^{-1}$ and $e = 1.60 \times 10^{-19}\ C$.)

Q1 An alpha particle (charge +2e) was deflected while passing through thin gold foil. The alpha particle passed within 5×10^{-12} m of a gold nucleus (charge +79e). What was the magnitude and direction of the electrostatic force experienced by the alpha particle? [4 marks]

Q2 (a) Two parallel plates are connected to a 1500 V dc supply, and separated by an air gap of 4.5 mm. What is the electric field strength between the plates? State the direction of the field. [3 marks]

(b) The plates are now pulled further apart so that the distance between them is doubled. The electric field strength remains the same. What is the new voltage between the plates? [2 marks]

Electric fields — one way to roast beef...

At least you get a choice here — uniform or radial, positive or negative, attractive or repulsive, chocolate or strawberry...

Electric Fields

Electric Potential is *Potential Energy per Unit Charge*

All points in an **electric field** have an **electric potential, *V***. This is the electric **potential energy** that a **unit positive charge** (+ 1 C) would have at that point. The **electric potential** of a point depends on **how far** it is from the **charge** creating the **electric field** and the **size** of that charge.

In a **radial field, electric potential** is given by:

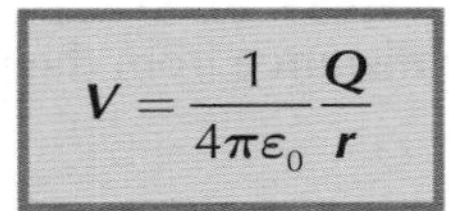

$$V = \frac{1}{4\pi\varepsilon_0}\frac{Q}{r}$$

where ***V*** is **electric potential** (V), ***Q*** is the size of the **charge** (C) and ***r*** is the **distance** from the charge (m)

1) The **sign** of ***V*** depends on the charge ***Q*** — i.e. ***V*** is **positive** when ***Q*** is positive and the force is **repulsive**, and **negative** when ***Q*** is negative and the force is **attractive**.
2) The **absolute magnitude** of ***V*** is **greatest** on the **surface of the charge**, and **decreases** as the **distance** from the charge **increases** — ***V*** will be **zero** at an **infinite distance** from the charge.

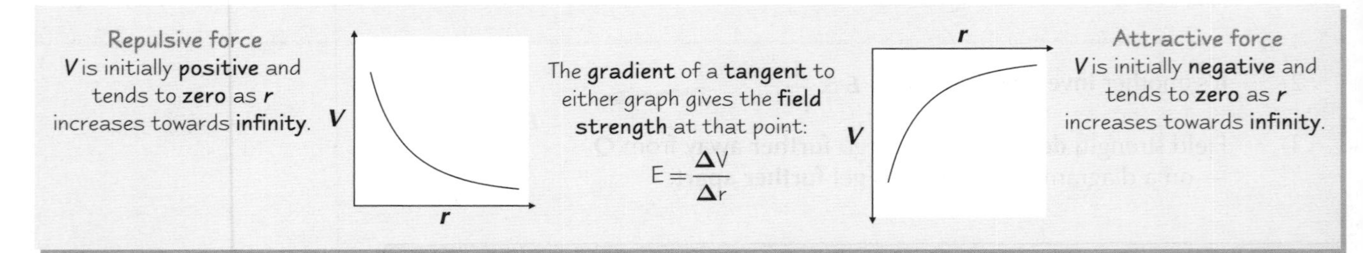

Electric Potential Difference is the *Energy Needed* to *Move* a *Unit Charge*

If **two points** in an **electric field** have different potential, then there is an **electric potential difference** between them. To **move a charge** across a **potential difference** (i.e. from one electric potential to another) you need to use **energy**.

The **amount of energy** you need (or the **work done**) depends on the **size** of the **charge** you're moving and the size of the **potential difference** you want to move it across:

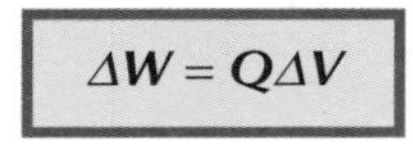

$$\Delta W = Q\Delta V$$

where ΔW is the work done (J), Q is the charge being moved (C) and ΔV is the electric potential difference (V).

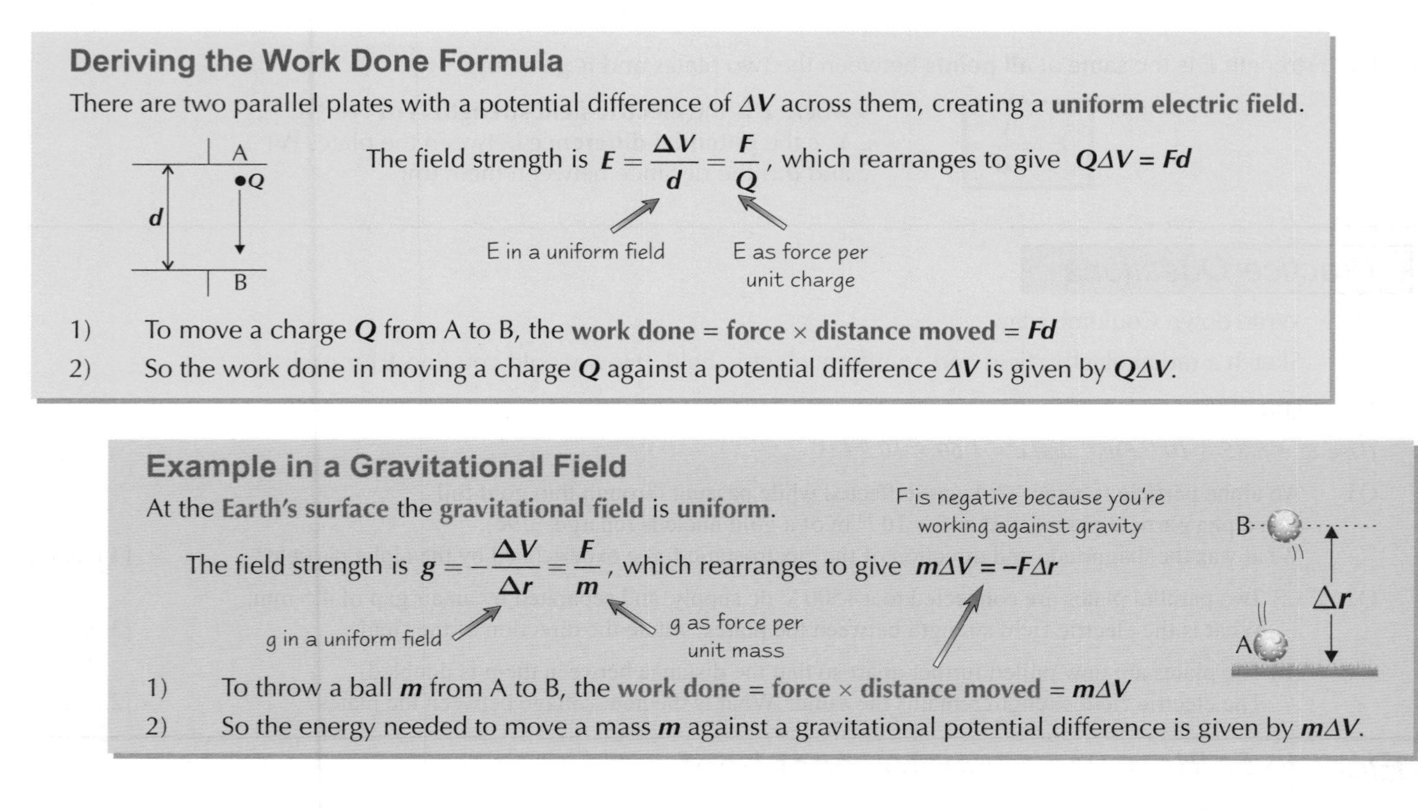

Deriving the Work Done Formula

There are two parallel plates with a potential difference of ΔV across them, creating a **uniform electric field**.

The field strength is $E = \frac{\Delta V}{d} = \frac{F}{Q}$, which rearranges to give $Q\Delta V = Fd$

1) To move a charge ***Q*** from A to B, the **work done = force × distance moved** = ***Fd***
2) So the work done in moving a charge ***Q*** against a potential difference ΔV is given by $Q\Delta V$.

Example in a Gravitational Field

At the **Earth's surface** the **gravitational field** is **uniform**.

The field strength is $g = -\frac{\Delta V}{\Delta r} = \frac{F}{m}$, which rearranges to give $m\Delta V = -F\Delta r$

1) To throw a ball ***m*** from A to B, the **work done = force × distance moved** = $m\Delta V$
2) So the energy needed to move a mass ***m*** against a gravitational potential difference is given by $m\Delta V$.

Electric Fields

There are Similarities between Gravitational and Electric Fields...

If a lot of the stuff on the previous couple of pages sounded strangely familiar it could be because it's very similar to the stuff on gravitational fields (or it could be because you've learnt it before — this is a revision book after all). Anyway, there are **four** big **similarities** between **electric** and **gravitational fields** that you need to know — read on.

1)	Gravitational field strength, g, is **force** per **unit mass**.	Electric field strength, E, is **force** per **unit positive charge**.
2)	Newton's law of gravitation for the **force** between two point masses is an **inverse square law**. $F \propto \frac{1}{r^2}$	Coulomb's law for the electric **force** between two point charges is also an **inverse square law**. $F \propto \frac{1}{r^2}$
3)	The **field lines** for a point mass... (m)	The **field lines** for a **negative** point charge... (+q, -Q)
4)	Gravitational potential, V, is **potential energy** per **unit mass** and is **zero** at **infinity**.	Electric potential, V, is **potential energy** per **unit positive charge** and is **zero** at **infinity**.

... and Three Differences too

Gravitational and electric fields aren't all the same — you need to know the **three main differences**:

1) Gravitational forces are always **attractive**. Electric forces can be either **attractive** or **repulsive**.
2) Objects can be **shielded** from **electric** fields, but not from gravitational fields.
3) The size of an **electric** force depends on the **medium** between the charges, e.g. plastic or air. For gravitational forces, this makes no difference.

Practice Questions

Q1 What is meant by 'electric potential'? How would you find the electric potential in a radial field?

Q2 Sketch a graph of electric potential against distance for an attractive and a repulsive charge.

Q3 What is 'potential difference'? What is the work done when a charge is moved through a potential difference?

Q4 Describe three similarities and three differences between gravitational and electric fields.

Exam Questions

Q1 Compare the magnitude and direction of the gravitational and electric forces between two electrons that are 8×10^{-10} m apart.
(Use $\boldsymbol{m_e} = 9.11 \times 10^{-31}$ kg, $\boldsymbol{e} = 1.60 \times 10^{-19}$ C, $\boldsymbol{\varepsilon_0} = 8.85 \times 10^{-12}$ Fm^{-1} and $\boldsymbol{G} = 6.67 \times 10^{-11}$ Nm^{2}kg^{-2}.) [3 marks]

Q2 A negatively charged oil drop is held stationary between two charged plates. The plates are 3 cm apart vertically, and have an electric potential difference of 5000 V across them.

5000 V

3 cm

(a) The oil drop has a mass of 1.5×10^{-14} kg. Calculate the size of its charge. [3 marks]

(b) If the polarity of the charged plates was reversed, what would happen to the oil drop? [1 mark]

I prefer gravitational fields — electric fields are repulsive...

Revising fields is a bit like a buy-one-get-one-free sale — you learn all about gravitational fields and they throw electric fields in for free. You just have to remember to change your ms for Qs and your Gs for $1/4\pi\varepsilon_0$s... okay, so it's not quite a BOGOF sale. Maybe more like a buy-one-get-one-half-price sale... anyway, you get the point — go learn some stuff.

Capacitors

Capacitors are things that store electrical charge — like a charge bucket. The capacitance of one of these things tells you how much charge the bucket can hold. Sounds simple enough... ha... ha, ha, ha...

Capacitance is Defined as the Amount of Charge Stored per Volt

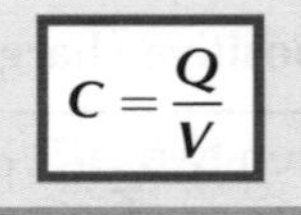

$$C = \frac{Q}{V}$$

where Q is the **charge** in coulombs, V is the **potential difference** in volts and C is the **capacitance in farads (F) — 1 farad = 1 C V^{-1}.**

A farad is a **huge** unit so you'll usually see capacitances expressed in terms of:

μF — microfarads (× 10^{-6})
nF — nanofarads (× 10^{-9})
pF — picofarads (× 10^{-12})

You can Investigate the Charge Stored by a Capacitor Experimentally

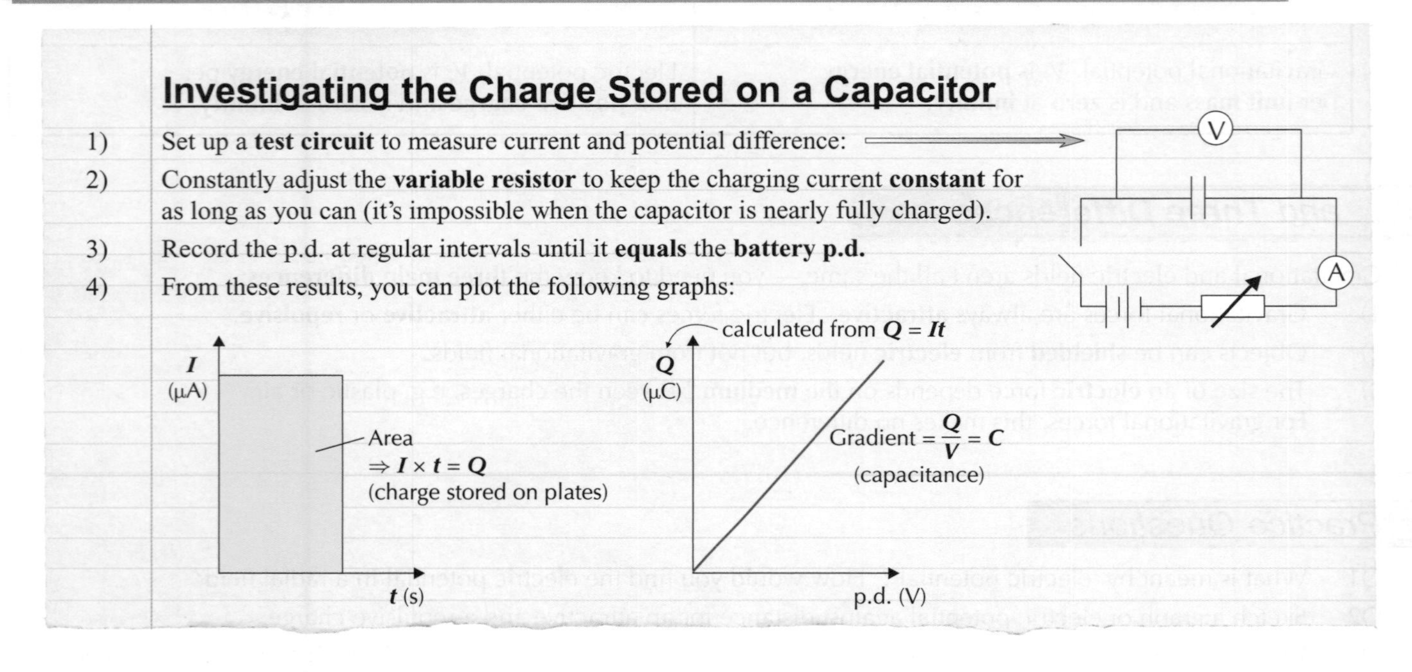

Investigating the Charge Stored on a Capacitor

1) Set up a **test circuit** to measure current and potential difference:
2) Constantly adjust the **variable resistor** to keep the charging current **constant** for as long as you can (it's impossible when the capacitor is nearly fully charged).
3) Record the p.d. at regular intervals until it **equals** the **battery p.d.**
4) From these results, you can plot the following graphs:

Capacitors Store Energy

1) When **charge** builds up on the plates of the **capacitor**, **electrical energy** is **stored** by the capacitor.
2) You can find the **energy stored** in a capacitor from the **area** under a **graph** of **potential difference** against **charge stored** on the capacitor.

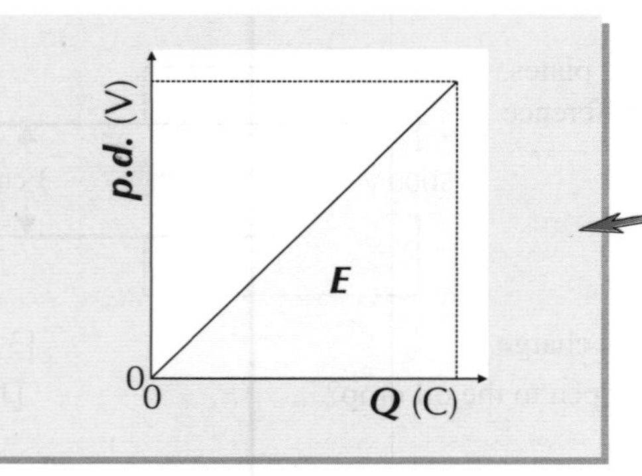

3) The p.d. across the capacitor is **proportional** to the charge stored on it (see above), so the graph will be a **straight line** through the origin. On this graph, the **energy stored** is given by the **yellow triangle**.

Jane had heard there was energy stored on plates.

4) **Area of triangle = ½ × base × height**, so the energy stored by the capacitor is:

$$E = \frac{1}{2}QV$$

You need to remember where this equation comes from.

Capacitors

There are **Three** Expressions for the **Energy Stored** by a Capacitor

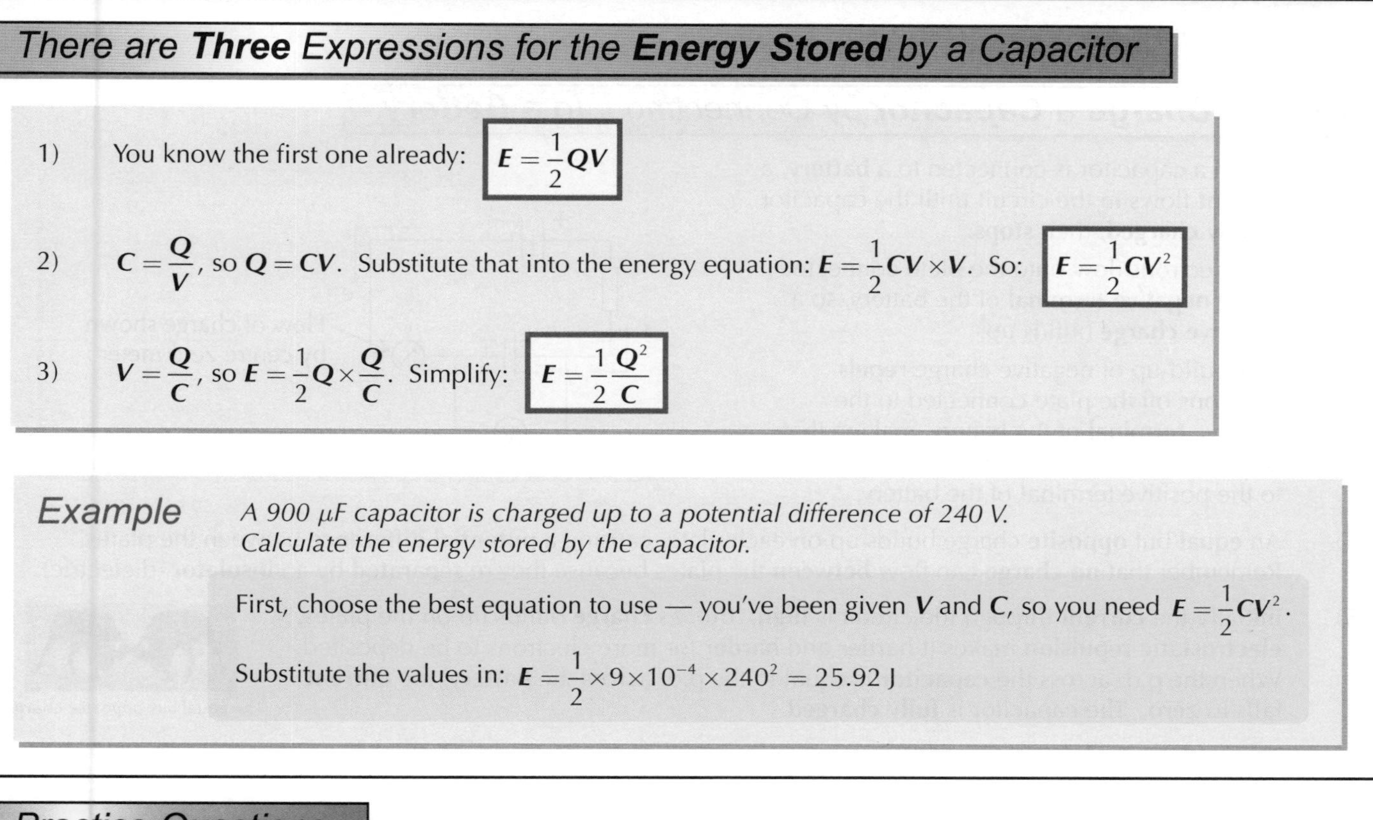

1) You know the first one already: $E = \frac{1}{2}QV$

2) $C = \frac{Q}{V}$, so $Q = CV$. Substitute that into the energy equation: $E = \frac{1}{2}CV \times V$. So: $E = \frac{1}{2}CV^2$

3) $V = \frac{Q}{C}$, so $E = \frac{1}{2}Q \times \frac{Q}{C}$. Simplify: $E = \frac{1}{2}\frac{Q^2}{C}$

Example — *A 900 μF capacitor is charged up to a potential difference of 240 V. Calculate the energy stored by the capacitor.*

First, choose the best equation to use — you've been given V and C, so you need $E = \frac{1}{2}CV^2$.

Substitute the values in: $E = \frac{1}{2} \times 9 \times 10^{-4} \times 240^2 = 25.92$ J

Practice Questions

Q1 Define capacitance.

Q2 What is the relationship between charge, voltage and capacitance?

Q3 Write the following in standard form: a) 220 μF b) 1000 pF c) 470 nF.

Exam Questions

Q1 The current and potential difference of a test circuit were measured as a capacitor was charged. The graphs below were plotted from the recorded data.

(a) Explain what is meant by the term 'capacitance'. [1 mark]

(b) Calculate: (i) the capacitance of the capacitor. [2 marks]

(ii) the charge stored on the plates of the capacitor. [2 marks]

Q2 (a) Sketch a circuit that could be used to find the energy stored on a capacitor. [2 marks]

(b) Explain how you would use this circuit to find the energy stored. [3 marks]

Q3 A 500 mF capacitor is fully charged up from a 12 V supply.

(a) Calculate the total energy stored by the capacitor. [2 marks]

(b) Calculate the charge stored by the capacitor. [2 marks]

Capacitance — fun, it's not...

Capacitors are really useful in the real world. Pick an appliance, any appliance, and it'll probably have a capacitor or several. If I'm being honest, though, the only saving grace of these pages for me is that they're not especially hard...

Charging and Discharging

Charging and discharging — sounds painful...

You can **Charge** a **Capacitor** by Connecting it to a **Battery**

1) When a capacitor is connected to a **battery**, a **current** flows in the circuit until the capacitor is **fully charged**, then **stops**.
2) The electrons flow onto the plate connected to the **negative terminal** of the battery, so a **negative charge** builds up.
3) This build-up of negative charge **repels** electrons off the plate connected to the **positive terminal** of the battery, making that plate positive. These electrons are attracted to the positive terminal of the battery.

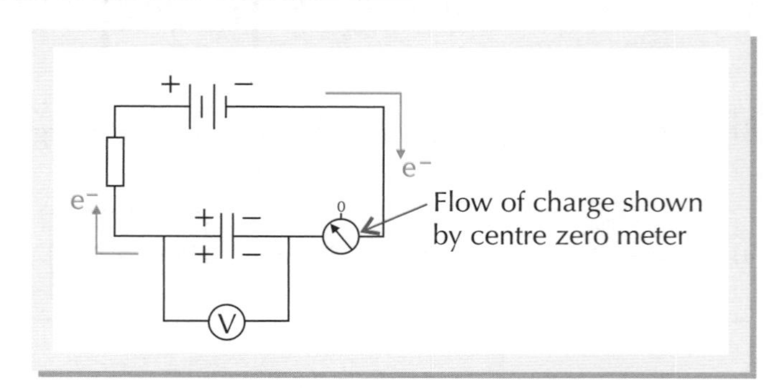

4) An **equal** but **opposite** charge builds up on each plate, causing a **potential difference** between the plates. Remember that **no charge** can flow **between** the plates because they're **separated** by an **insulator** (dielectric).
5) Initially the **current** through the circuit is **high**. But, as **charge** builds up on the plates, **electrostatic repulsion** makes it **harder** and **harder** for more electrons to be deposited. When the p.d. across the **capacitor** is equal to the p.d. across the **battery**, the **current** falls to **zero**. The capacitor is **fully charged**.

an equal but opposite charge

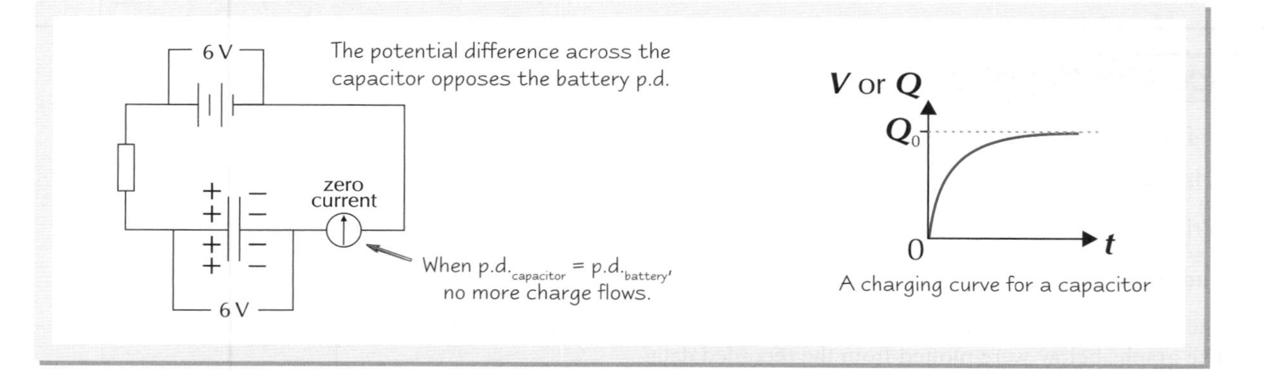

A charging curve for a capacitor

To **Discharge** a Capacitor, **Take Out** the **Battery** and **Reconnect** the **Circuit**

1) When a **charged capacitor** is connected across a **resistor**, the p.d. drives a **current** through the circuit.
2) This current flows in the **opposite direction** from the **charging current**.
3) If you connect a **voltage sensor** attached to a **datalogger** across the **capacitor**, you can plot a **discharge curve** for the capacitor (like the two below). The curve shows how the **voltage changes** as the **capacitor discharges**.
4) The capacitor is **fully discharged** when the **p.d.** across the plates and the **current** in the circuit are both **zero**.

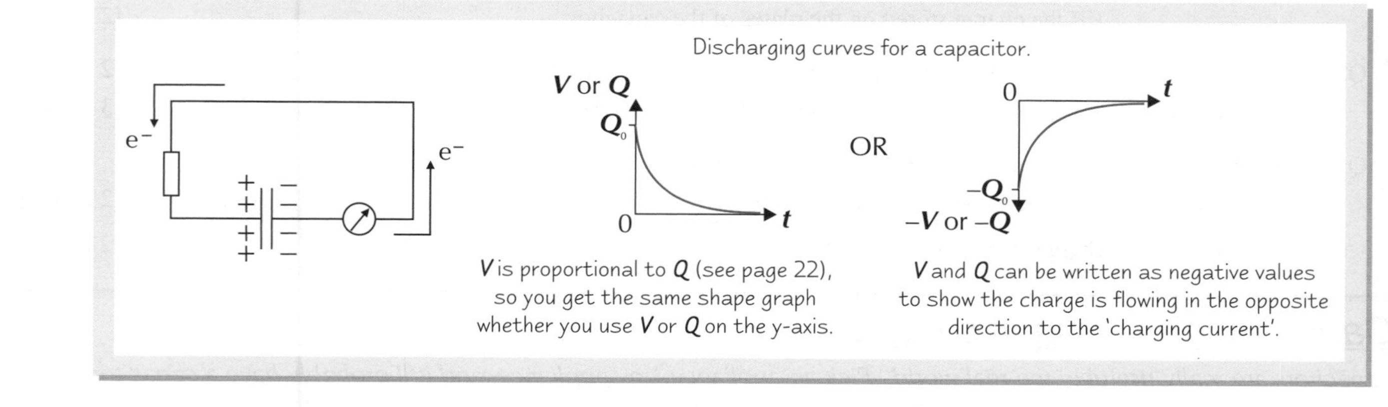

V is proportional to *Q* (see page 22), so you get the same shape graph whether you use *V* or *Q* on the y-axis.

V and *Q* can be written as negative values to show the charge is flowing in the opposite direction to the 'charging current'.

Charging and Discharging

The Time Taken to Charge or Discharge Depends on Two Factors

The **time** it takes to charge or discharge a capacitor depends on:

1) The **capacitance** of the capacitor (***C***). This affects the amount of **charge** that can be transferred at a given **voltage**.
2) The **resistance** of the circuit (***R***). This affects the **current** in the circuit.

The Charge on a Capacitor Decreases Exponentially

1) When a capacitor is **discharging**, the amount of **charge** left on the plates falls **exponentially with time**.
2) That means it always takes the **same length of time** for the charge to **halve**, no matter **how much charge** you start with — like radioactive decay (see p. 38).

The charge left on the plates of a capacitor discharging from full is given by the equation:

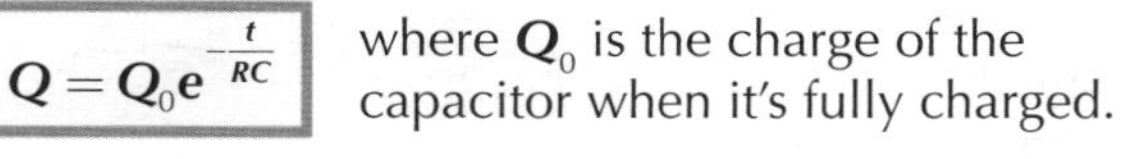

$$Q = Q_0 e^{-\frac{t}{RC}}$$

where Q_0 is the charge of the capacitor when it's fully charged.

Charging: $V = V_0 - V_0 e^{-\frac{t}{RC}}$

Discharging: $V = V_0 e^{-\frac{t}{RC}}$

3) The graphs of ***V*** against ***t*** for charging and discharging are also **exponential**.
4) ***I*** for a charging or discharging capacitor **always decreases exponentially** — the graph of ***I*** against ***t*** is the **same shape** as the graph of ***V*** against ***t*** for a discharging capacitor.

Time Constant τ = RC

τ is the Greek letter 'tau'

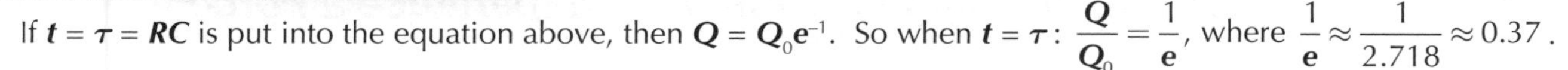

If $t = \tau = RC$ is put into the equation above, then $Q = Q_0 e^{-1}$. So when $t = \tau$: $\frac{Q}{Q_0} = \frac{1}{e}$, where $\frac{1}{e} \approx \frac{1}{2.718} \approx 0.37$.

1) So τ, the **time constant**, is the time taken for the charge on a discharging capacitor (*Q*) to **fall** to **37%** of Q_0.
2) It's also the time taken for the charge of a charging capacitor to **rise** to **63%** of Q_0.
3) The **larger** the **resistance** in series with the capacitor, the **longer it takes** to charge or discharge.
4) In practice, the time taken for a capacitor to charge or discharge **fully** is taken to be about 5***RC***.

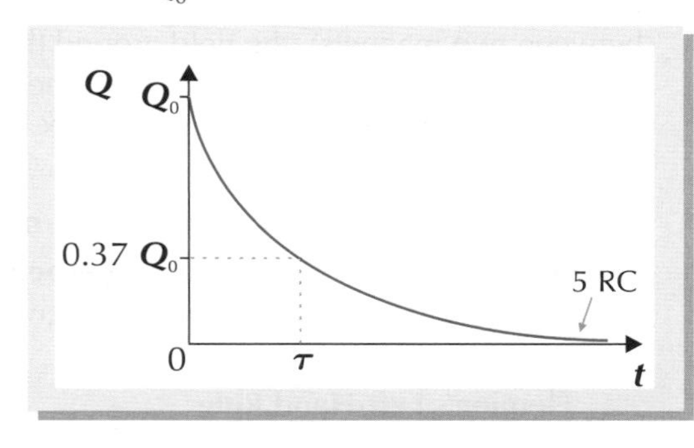

Practice Questions

Q1 Sketch graphs to show the variation of p.d. across the plates of a capacitor with time for:
a) charging a capacitor, b) discharging a capacitor.

Q2 What two factors affect the rate of charge of a capacitor?

Exam Question

Q1 A 250 μF capacitor is fully charged from a 6 V battery and then discharged through a 1 kΩ resistor.

(a) Calculate the time taken for the charge on the capacitor to fall to 37% of its original value. [2 marks]

(b) Calculate the percentage of the total charge remaining on the capacitor after 0.7s. [3 marks]

(c) The charging voltage is increased to 12 V. Explain the effect this has on:

i) the total charge stored ii) the capacitance of the capacitor iii) the time taken to fully charge [3 marks]

An analogy — consider the lowly bike pump...

A good way to think of the charging process is like pumping air into a bike tyre. To start with, the air goes in easily, but as the pressure in the tyre increases, it gets harder and harder to squeeze any more air in. The tyre's 'full' when the pressure of the air in the tyre equals the pressure of the pump. The analogy works just as well for discharging...

Magnetic Fields

Magnetic fields — making pretty patterns with iron filings before spending an age trying to pick them off the magnet.

A Magnetic Field is a Region Where a Force is Exerted on Magnetic Materials

1) Magnetic fields can be represented by **field lines**.
2) Field lines go from **north to south**.
3) The **closer** together the lines, the **stronger** the field.

N S N S N N

At a neutral point magnetic fields cancel out.

There is a Magnetic Field Around a Wire Carrying Electric Current

When **current** flows in a **wire**, a **magnetic field** is induced around the wire.

1) The **field lines** are **concentric circles** centred on the wire.
2) The **direction** of a magnetic **field** around a current-carrying wire can be worked out with the **right-hand rule**.
3) If you loop the wire into a **coil**, the field is **doughnut shaped**, while lots of coils (a **solenoid**) form a **field** like a **bar magnet**.

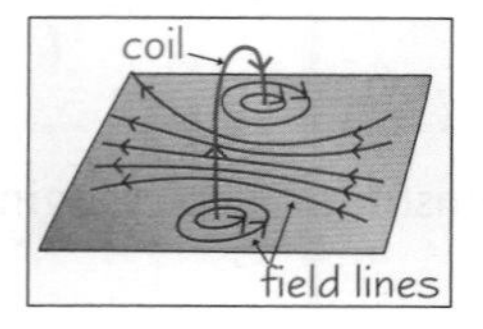

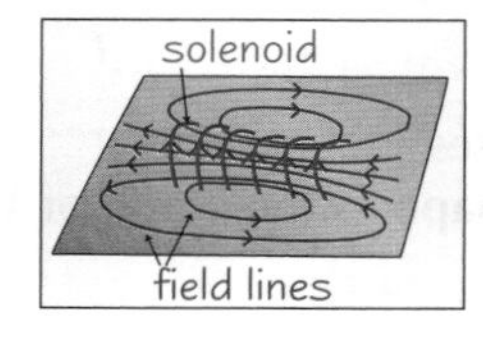

RIGHT-HAND RULE

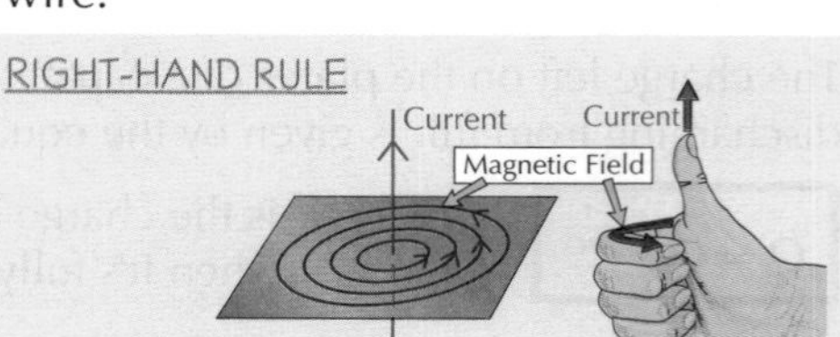

1) Stick your right thumb up, like you're hitching a lift.
2) If your thumb points in the direction of the current...
3) ...your curled fingers point in the direction of the field.

A Wire Carrying a Current in a Magnetic Field will Experience a Force

1) If you put a **current-carrying wire** into an **external** magnetic field (e.g. between two magnets), the field around the wire and the field from the magnets **interact**. The field lines from the magnet **contract** to form a **'stretched catapult'** effect where the flux lines are closer together.
2) This causes a **force** on the wire.
3) If the current is **parallel** to the flux lines, **no force** acts.
4) The **direction** of the force is always **perpendicular** to both the **current** direction and the **magnetic field** — it's given by **Fleming's left-hand rule**:

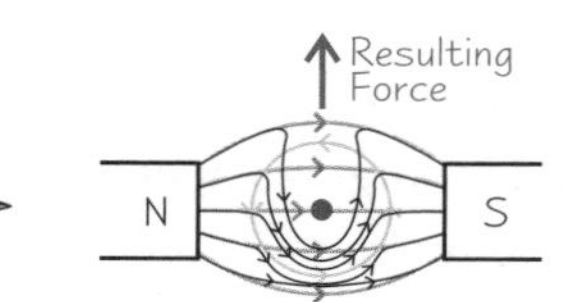

Fleming's Left-Hand Rule

The First finger points in the direction of the uniform magnetic Field, the seCond finger points in the direction of the conventional Current. Then your thuMb points in the direction of the force (in which Motion takes place).

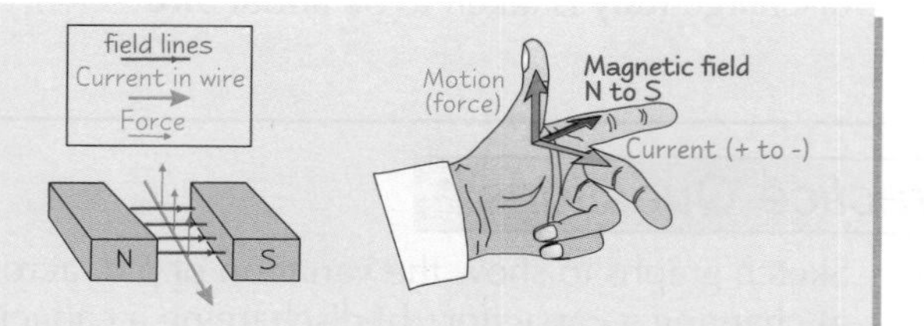

The Force on a Wire is Proportional to the Magnetic Field Strength

1) The size of the **force**, ***F***, on a current-carrying wire at right-angles to a magnetic field is proportional to the **current**, ***I***, the **length of wire** in the field, ***l***, and the **strength of the magnetic field**, ***B***. This gives the equation: $F = BIl$
2) In this equation, the **magnetic field strength**, ***B***, is defined as:

The **force** on **one metre** of wire carrying a **current** of **one amp** at **right angles** to the **magnetic field**.

3) **Magnetic field strength** is also called **flux density** and it's measured in **teslas**, **T**. $1 \text{ tesla} = \frac{\text{Wb}}{\text{m}^2}$
4) Magnetic field strength is a **vector** quantity with both a **direction** and **magnitude**.

It helps to think of flux density as the number of flux lines (measured in webers (Wb), see p 28) per unit area.

Magnetic Fields

Forces Act on Charged Particles in Magnetic Fields

Electric current in a wire is caused by the **flow** of negatively **charged** electrons. These charged particles are affected by **magnetic fields** — so a current-carrying wire experiences a **force** in a magnetic field (see page 26).

1) The **force** on a **current-carrying wire** in a **magnetic field** perpendicular to the current is given by $F = BIl$.
2) Electric **current**, I, is the flow of **charge**, q, per unit **time**, t, i.e. $q = It$.
3) A charged particle which moves a **distance** l in **time** t has a **velocity**, $v = l/t$.
4) Putting all these equations **together** (with a spot of rearranging) gives the **force** acting on a **single charged particle moving through a magnetic field**:

In many exam questions, q is the size of the charge on the electron, which is 1.60×10^{-19} coulomb.

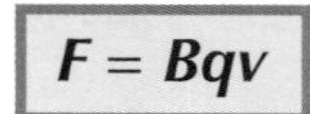

$$F = Bqv$$

where B is the **magnetic field strength** (T), q is the **charge** on the particle (C), and v is its **velocity** (ms^{-1})

Example

What is the force acting on an electron travelling at 2×10^4 ms^{-1} through a uniform magnetic field of strength 2 T? (The magnitude of the charge on an electron is 1.6×10^{-19} C.)

Use $F = Bqv$:

$F = 2 \times 1.6 \times 10^{-19} \times 2 \times 10^4 = \mathbf{6.4 \times 10^{-15}}$ **N**

Charged Particles in a Magnetic Field are Deflected in a Circular Path

1) By **Fleming's left-hand rule** the force on a **moving charge** in a magnetic field is always **perpendicular** to its **direction of travel**.
2) Mathematically, that is the condition for **circular** motion.
3) This effect is used in **particle accelerators** such as **cyclotrons** and **synchrotrons**, which use **magnetic fields** to accelerate particles to very **high energies** along circular paths.
4) The **radius of curvature** of the **path** of a charged particle moving through a magnetic field gives you information about the particle's **charge** and **mass** — this means you can **identify different particles** by studying how they're **deflected**.

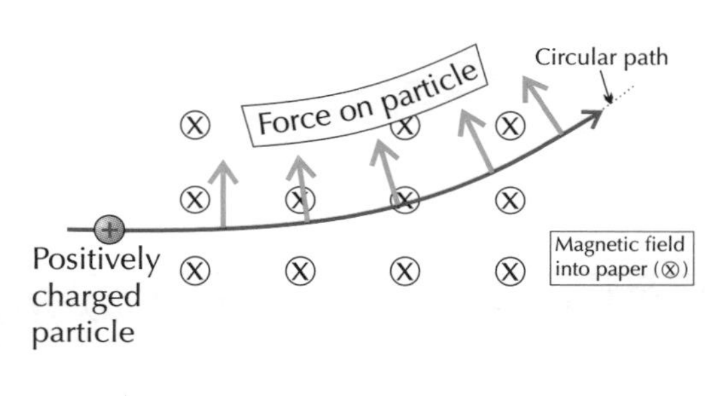

Practice Questions

Q1 Describe why a current-carrying wire at right angles to an external magnetic field will experience a force.

Q2 Write down the equation you would use to find the force on a current-carrying wire.

Q3 Sketch the magnetic fields around a long straight current-carrying wire. Show the direction of the current and magnetic field.

Q4 A copper bar can roll freely on two copper supports, as shown in the diagram. When current is applied in the direction shown, which way will the bar roll?

Horseshoe magnet

N

S

Copper bar

Exam Question

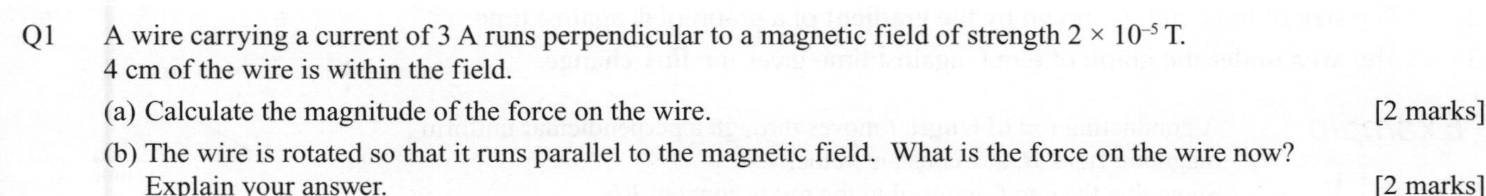

Q1 A wire carrying a current of 3 A runs perpendicular to a magnetic field of strength 2×10^{-5} T. 4 cm of the wire is within the field.

(a) Calculate the magnitude of the force on the wire. [2 marks]

(b) The wire is rotated so that it runs parallel to the magnetic field. What is the force on the wire now? Explain your answer. [2 marks]

I revised the right-hand rule by the A69 and ended up in Newcastle...

Fleming's left-hand rule is the key to this section — so make sure you know how to use it and understand what it all means. Remember that the direction of the magnetic field is from N to S, and that the current is from +ve to –ve — this is as important as using the correct hand. You need to get those right or it'll all go to pot...

Electromagnetic Induction

Think of the Magnetic Flux as the Total Number of Field Lines

1) **Magnetic flux density**, ***B***, is a measure of the **strength** of the magnetic field **per unit area**.
2) The total **magnetic flux**, ϕ, passing through an **area**, ***A***, perpendicular to a **magnetic field**, ***B***, is defined as:

$$\phi = BA$$

where ϕ is **magnetic flux** (Wb), ***B*** is **magnetic field strength** (T) and ***A*** is **area** (m^2)

3) When you move a **coil** in a magnetic field, the size of the e.m.f. induced depends on the **magnetic flux** passing through the coil, ϕ, and the **number of turns** on the coil. The product of these is called the **flux linkage**, Φ.

For a coil of ***N*** turns perpendicular to B, the flux linkage is given by:

$$\Phi = N\phi = BAN$$

ϕ is the little Greek letter 'phi', and Φ is a capital 'phi'. The unit of both ϕ and Φ is the **weber, Wb**.

If the magnetic flux is **not perpendicular** to ***B***, you can find the **magnetic flux** using this equation:

$$\phi = BA\cos\theta$$

where θ is the **angle** between the **field** and the **normal** to the plane of the coil.

For a **coil** with ***N*** turns the **flux linkage** is:

$$\Phi = BAN\cos\theta$$

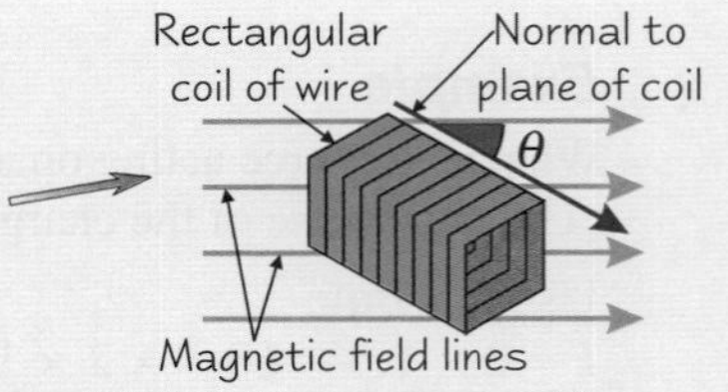

Charges Accumulate on a Conductor Moving Through a Magnetic Field

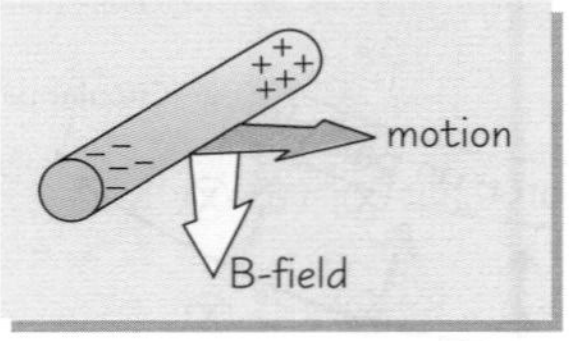

1) If a **conducting rod** moves through a magnetic field its **electrons** will experience a **force** (see p. 26), which means that they will **accumulate** at one end of the rod.
2) This **induces** an **e.m.f.** (**electromotive force**) across the ends of the rod.
3) If the rod is part of a complete **circuit**, then an induced **current** will **flow** through it — this is called **electromagnetic induction**.

Changes in Magnetic Flux Induce an Electromotive Force

1) An **electromotive force** (**e.m.f.**) is **induced** when there is **relative motion** between a **conductor** and a **magnet**.
2) The **conductor** can **move** and the **magnetic field** stay **still** or the **other way round** — you get an e.m.f. either way.
3) **Flux cutting** always induces e.m.f. but will only **induce** a **current** if the **circuit** is complete.
4) **Flux linking** is when an e.m.f. is induced by **changing** the **magnitude** or **direction** of the **magnetic flux** (e.g. caused by an **alternating current** electromagnet).

A **change in flux** of **one weber per second** will induce an **electromotive force** of **1 volt** in a loop of wire.

These Results are Summed up by Faraday's Law...

FARADAY'S LAW: The **induced e.m.f.** is **directly proportional** to the **rate of change of flux linkage**.

1) **Faraday's law** can be written as: $\text{Induced e.m.f.} = \dfrac{\text{flux change}}{\text{time taken}} = N\dfrac{\Delta\phi}{\Delta t}$
2) The **size** of the e.m.f. is shown by the **gradient** of a graph of Φ against time.
3) The **area under** the graph of e.m.f. against time gives the **flux change**.

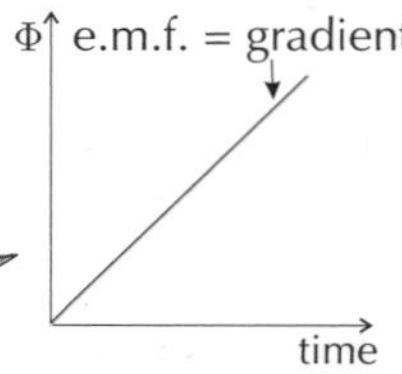

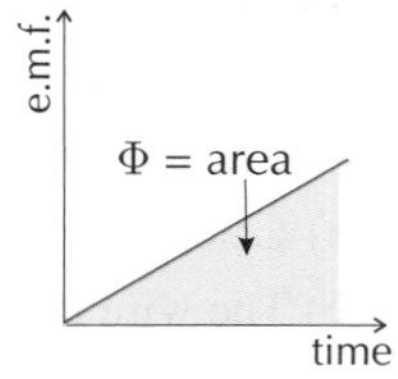

You might be asked to find the e.m.f. induced by the Earth's magnetic field across the wingspan of a plane — just think of it as a moving rod and use the equation as usual.

Example

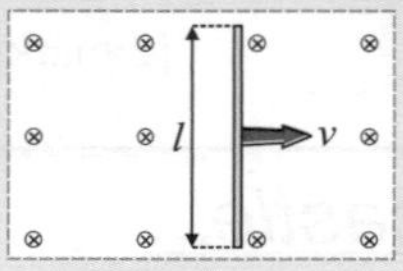

A conducting rod of **length *l*** moves through a perpendicular uniform magnetic field, ***B***, at a constant velocity, ***v***.
Show that the e.m.f. induced in the rod is equal to ***Blv***.

The distance travelled by the rod is $v\Delta t$.
So the area of flux it cuts is $lv\Delta t$, and total magnetic flux is $\phi = Blv\Delta t$.

Faraday's law states that $\text{e.m.f} = \dfrac{\Delta\phi}{\Delta t}$, so $\text{e.m.f} = \dfrac{Blv\Delta t}{\Delta t} = Blv$

Electromagnetic Induction

The **Direction** of the **Induced E.m.f.** and **Current** are given by **Lenz's Law**...

LENZ'S LAW: The **induced e.m.f.** is always in such a **direction** as to **oppose** the **change** that caused it.

1) **Lenz's law** and **Faraday's law** can be **combined** to give one formula that works for both:

$$\text{Induced e.m.f.} = -N\frac{\Delta\phi}{\Delta t}$$

2) The **minus sign** shows the direction of the **induced e.m.f.**
3) The idea that an induced e.m.f. will **oppose** the change that caused it agrees with the principle of the **conservation of energy** — the **energy used** to pull a conductor through a magnetic field, against the **resistance** caused by magnetic **attraction**, is what **produces** the **induced current**.
4) **Lenz's law** can be used to find the **direction** of an **induced e.m.f.** and **current** in a conductor travelling at right angles to a magnetic field.

1) **Lenz's law** says that the **induced e.m.f.** will produce a force that **opposes** the motion of the conductor — in other words a **resistance**.
2) Using **Fleming's left-hand rule** (see p.26), point your thumb in the direction of the force of **resistance** — which is in the **opposite direction** to the motion of the conductor.
3) Your **second finger** will now give you the direction of the **induced e.m.f.**
4) If the conductor is **connected** as part of a **circuit**, a current will be induced in the **same direction** as the induced e.m.f.

Practice Questions

Q1 What is the difference between magnetic flux density, magnetic flux and magnetic flux linkage?

Q2 A coil consists of N turns, each of area A. What is its flux linkage:
a) if it is perpendicular to a uniform magnetic field, b) if it is at an angle to a uniform magnetic field?

Q3 State Faraday's law.

Q4 State Lenz's law.

Q5 Explain how to find the direction of an induced e.m.f. in a copper bar moving at right angles to a magnetic field.

Exam Questions

Q1 A coil of area 0.23 m^2 is placed at right angles to a magnetic field of 2×10^{-3} T.
(a) Calculate the magnetic flux passing through the coil. [2 marks]
(b) If the coil has 150 turns, what is the magnetic flux linkage in the coil? [2 marks]
(c) Over a period of 2.5 seconds the magnetic field is reduced uniformly to 1.5×10^{-3} T.
Calculate the size of the e.m.f. induced across the ends of the coil. [3 marks]

Q2 An aeroplane with a wingspan of 30 m flies at a speed of 100 ms^{-1} perpendicular to the Earth's magnetic field, as shown.
The Earth's magnetic field at the aeroplane's location is 60×10^{-6} T.
(a) Calculate the induced e.m.f. between the wing tips. [2 marks]
(b) Complete the diagram to show the direction of the induced e.m.f. between the wing-tips. [1 mark]

Beware — physics can induce extreme confusion...

Make sure you know the difference between flux and flux linkage, and that you can calculate both. Then all you need to learn is that the induced e.m.f. is proportional to minus the rate of change of flux linkage — and that's it. Remember when you're using Fleming's left-hand rule to work out the direction of the induced e.m.f. that you need to point your thumb in the opposite direction to the direction the conductor is moving... Phew, hope you got all that — I need a cuppa to recover.

Transformers and Alternators

Turns out electromagnetic induction is quite useful in the real world — remember that place?

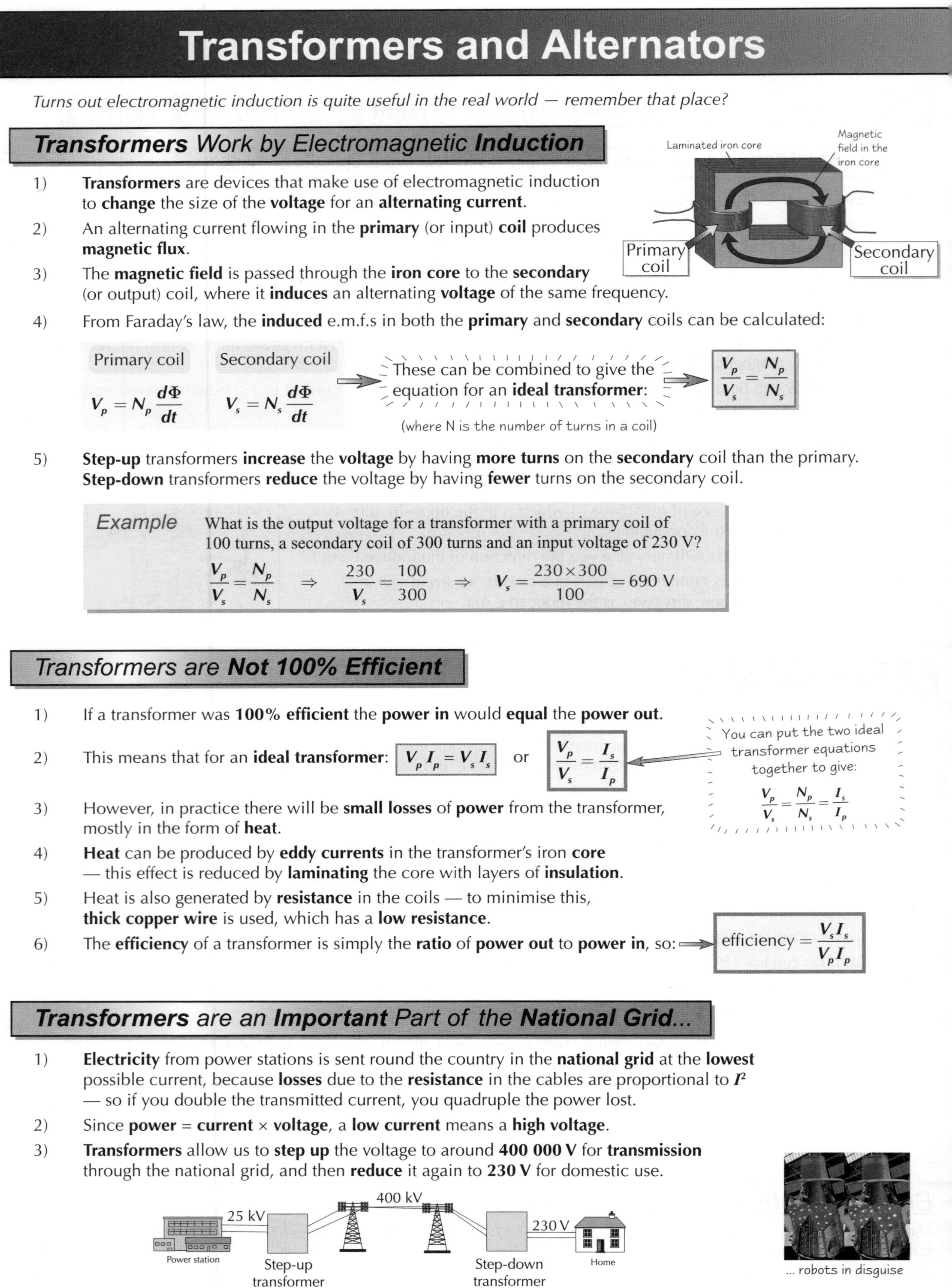

Transformers Work by Electromagnetic Induction

1) **Transformers** are devices that make use of electromagnetic induction to **change** the size of the **voltage** for an **alternating current**.
2) An alternating current flowing in the **primary** (or input) **coil** produces **magnetic flux**.
3) The **magnetic field** is passed through the **iron core** to the **secondary** (or output) coil, where it **induces** an alternating **voltage** of the same frequency.
4) From Faraday's law, the **induced** e.m.f.s in both the **primary** and **secondary** coils can be calculated:

Primary coil: $V_p = N_p \dfrac{d\Phi}{dt}$ Secondary coil: $V_s = N_s \dfrac{d\Phi}{dt}$

These can be combined to give the equation for an **ideal transformer**: $\dfrac{V_p}{V_s} = \dfrac{N_p}{N_s}$

(where N is the number of turns in a coil)

5) **Step-up** transformers **increase** the **voltage** by having **more turns** on the **secondary** coil than the primary. **Step-down** transformers **reduce** the voltage by having **fewer** turns on the secondary coil.

Example What is the output voltage for a transformer with a primary coil of 100 turns, a secondary coil of 300 turns and an input voltage of 230 V?

$$\frac{V_p}{V_s} = \frac{N_p}{N_s} \Rightarrow \frac{230}{V_s} = \frac{100}{300} \Rightarrow V_s = \frac{230 \times 300}{100} = 690\text{ V}$$

Transformers are Not 100% Efficient

1) If a transformer was **100% efficient** the **power in** would **equal** the **power out**.
2) This means that for an **ideal transformer**: $V_p I_p = V_s I_s$ or $\dfrac{V_p}{V_s} = \dfrac{I_s}{I_p}$

You can put the two ideal transformer equations together to give: $\dfrac{V_p}{V_s} = \dfrac{N_p}{N_s} = \dfrac{I_s}{I_p}$

3) However, in practice there will be **small losses** of **power** from the transformer, mostly in the form of **heat**.
4) **Heat** can be produced by **eddy currents** in the transformer's iron **core** — this effect is reduced by **laminating** the core with layers of **insulation**.
5) Heat is also generated by **resistance** in the coils — to minimise this, **thick copper wire** is used, which has a **low resistance**.
6) The **efficiency** of a transformer is simply the **ratio** of **power out** to **power in**, so: efficiency $= \dfrac{V_s I_s}{V_p I_p}$

Transformers are an Important Part of the National Grid...

1) **Electricity** from power stations is sent round the country in the **national grid** at the **lowest** possible current, because **losses** due to the **resistance** in the cables are proportional to I^2 — so if you double the transmitted current, you quadruple the power lost.
2) Since **power** = **current** × **voltage**, a **low current** means a **high voltage**.
3) **Transformers** allow us to **step up** the voltage to around **400 000 V** for **transmission** through the national grid, and then **reduce** it again to **230 V** for domestic use.

... robots in disguise

Transformers and Alternators

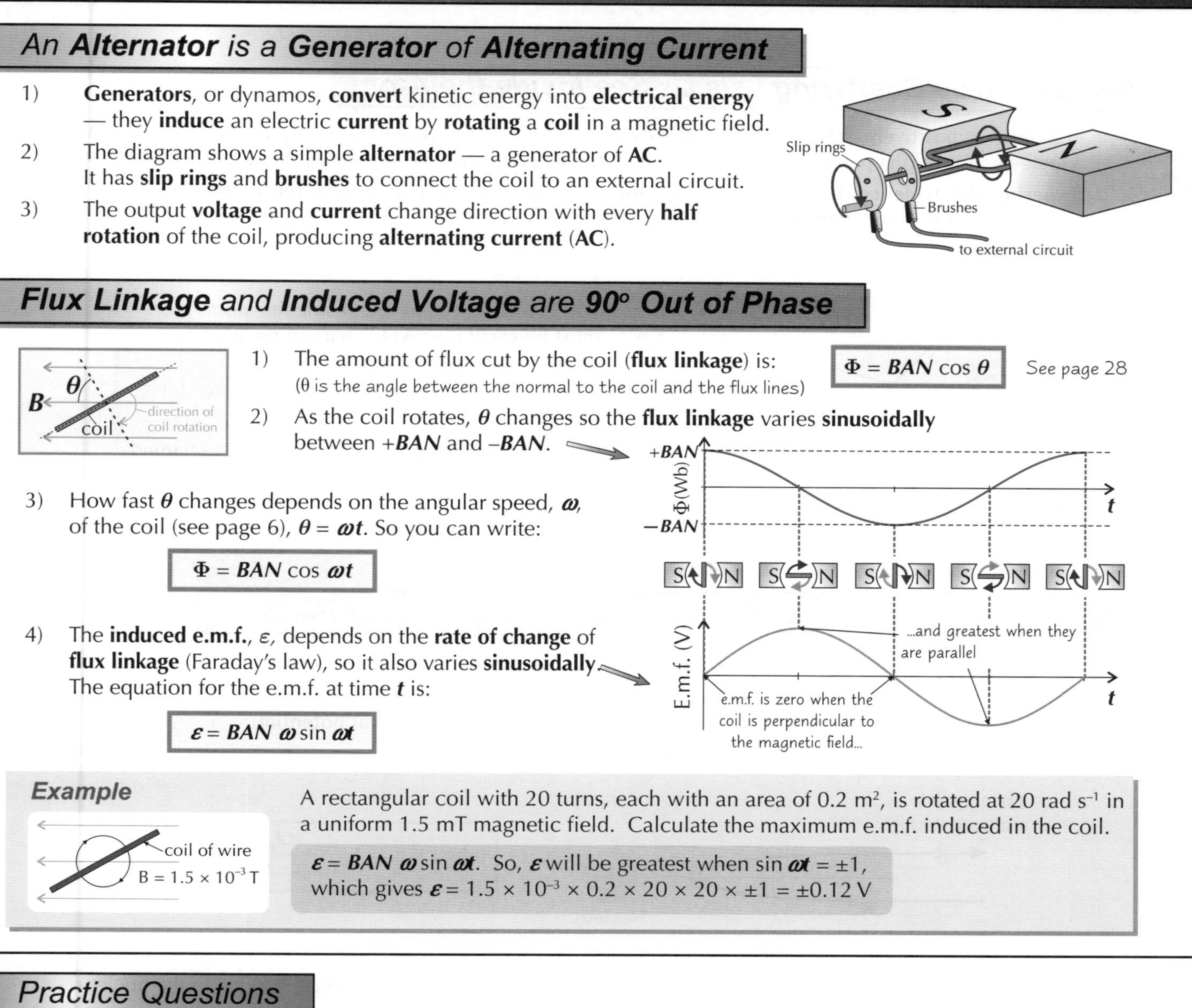

An Alternator is a Generator of Alternating Current

1) **Generators**, or dynamos, **convert** kinetic energy into **electrical energy** — they **induce** an electric **current** by **rotating** a **coil** in a magnetic field.
2) The diagram shows a simple **alternator** — a generator of **AC**. It has **slip rings** and **brushes** to connect the coil to an external circuit.
3) The output **voltage** and **current** change direction with every **half rotation** of the coil, producing **alternating current** (**AC**).

Flux Linkage and Induced Voltage are 90° Out of Phase

1) The amount of flux cut by the coil (**flux linkage**) is: $\Phi = BAN\cos\theta$ See page 28
(θ is the angle between the normal to the coil and the flux lines)
2) As the coil rotates, θ changes so the **flux linkage** varies **sinusoidally** between $+BAN$ and $-BAN$.
3) How fast θ changes depends on the angular speed, ω, of the coil (see page 6), $\theta = \omega t$. So you can write:

$$\Phi = BAN\cos\omega t$$

4) The **induced e.m.f.**, ε, depends on the **rate of change** of **flux linkage** (Faraday's law), so it also varies **sinusoidally**. The equation for the e.m.f. at time t is:

$$\varepsilon = BAN\omega\sin\omega t$$

Example

A rectangular coil with 20 turns, each with an area of 0.2 m², is rotated at 20 rad s⁻¹ in a uniform 1.5 mT magnetic field. Calculate the maximum e.m.f. induced in the coil.

$\varepsilon = BAN\omega\sin\omega t$. So, ε will be greatest when $\sin\omega t = \pm1$, which gives $\varepsilon = 1.5\times10^{-3}\times0.2\times20\times20\times\pm1 = \pm0.12$ V

Practice Questions

Q1 Draw a diagram of a simple transformer.

Q2 What is meant by a step-down transformer?

Q3 Show that flux linkage and induced e.m.f. are 90° out of phase.

Exam Questions

Q1 A transformer with 150 turns in the primary coil has an input voltage of 9 V.

(a) Calculate the number of turns needed in the secondary coil to step up the voltage to 45 V. [2 marks]

(b) The input current for the transformer is 1.5 A. Assuming the transformer is ideal, state the output current. [2 marks]

(c) Calculate the actual efficiency of the transformer given that the power output is measured as 10.8 W. [2 marks]

Q2 A 0.01 m² coil of 500 turns is rotated on an axis that is perpendicular to a magnetic field of 0.9 T.

(a) Find the flux linkage when the angle between its normal and the magnetic field is 60°. [2 marks]

(b) If the coil is rotated at an angular speed of 40π rad s⁻¹, what is the peak e.m.f. induced? [3 marks]

B 60° coil rotation

Arrrrrrrrggggggghhhhhh...

Breathe a sigh of relief, pat yourself on the back and make a brew — well done, you've reached the end of the section. That was pretty nasty stuff, but don't let all of those equations get you down — once you've learnt the main ones and can use them blindfolded, even the trickiest looking exam question will be a walk in the park...

Scattering to Determine Structure

By firing radiation at different materials, you can take a sneaky beaky at their internal structures...

Alpha Particle Scattering Lets Us See Inside the Atom

1) If a beam of **positively charged alpha particles** is directed at a thin gold film, most **pass straight through**. However, if an alpha particle is travelling straight towards, or close by, a nucleus, its path will be **deflected**.
2) **Experimental evidence** shows that some of these alpha particles are deflected by **more than 90°** — in other words they '**bounce**' back.

Rutherford Scattering

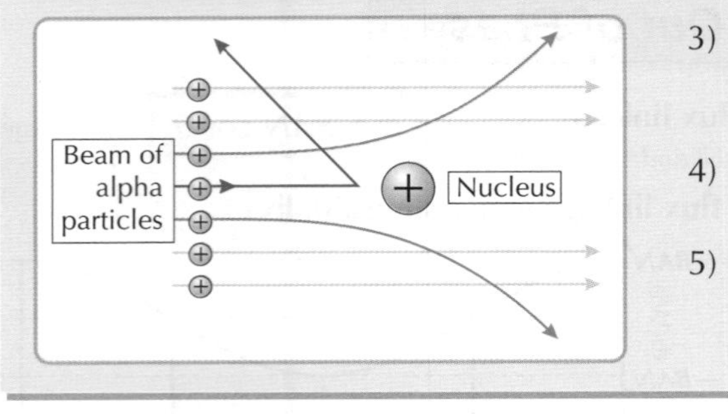

3) This evidence means that inside the atoms there must be **small positively charged nuclei**, which **repel** the passing alpha particles.
4) The nucleus must be **small** since very few alpha particles are deflected by much.
5) It must be **positive** to repel the positively charged alpha particles.

You can Estimate the Closest Approach of a Scattered Particle

1) When you fire an alpha particle at a gold nucleus, you know its **initial kinetic energy**.
2) An alpha particle that 'bounces back' and is deflected through 180° will have stopped a short distance from the nucleus. It does this at the point where its **electrical potential energy** (see p 20) **equals** its **initial kinetic energy**.

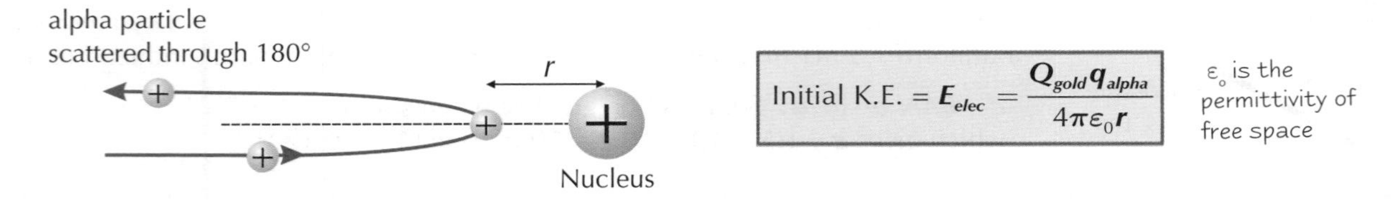

$$\text{Initial K.E.} = E_{elec} = \frac{Q_{gold}q_{alpha}}{4\pi\varepsilon_0 r}$$

ε_0 is the permittivity of free space

3) It's just conservation of energy — and you can use it to find how close the particle can get to the nucleus.
4) To find the charge of a nucleus you need to know the atom's **proton number, Z** — that tells you how many protons are in the nucleus (surprisingly).
A proton has a charge of **+e** (where e is the charge on an electron), so the charge of a nucleus must be **+Ze**.

Example An alpha particle with an initial kinetic energy of 6 MeV is fired at a gold nucleus. Find the closest approach of the alpha particle to the nucleus.

Initial particle energy = 6 MeV = 6×10^6 eV

Convert energy into joules: $6 \times 10^6 \times 1.6 \times 10^{-19} = 9.6 \times 10^{-13}$ J

So, electrical potential energy $= E_{elec} = \dfrac{Q_{gold}q_{alpha}}{4\pi\varepsilon_0 r} = 9.6 \times 10^{-13}$ J at closest approach.

Rearrange to get $r = \dfrac{(+79e)(+2e)}{4\pi\varepsilon_0(9.6 \times 10^{-13})}$

$$= \frac{2 \times 79 \times (1.6 \times 10^{-19})^2}{4\pi \times 8.9 \times 10^{-12} \times 9.6 \times 10^{-13}} = \mathbf{3.8 \times 10^{-14}\ m}$$

Scattering to Determine Structure

You can also use Electron Diffraction to Estimate Nuclear Diameter

1) **Electrons** are a type of particle called a **lepton**. Leptons **don't interact** with the **strong nuclear force** (whereas neutrons and alpha particles do). Because of this, electron diffraction is the **most accurate** method for getting a picture of a crystal's **atomic structure**.
2) Like other particles, electrons show **wave-particle duality** (see p 96) — so **electron beams** can be diffracted.
3) A beam of moving electrons has an associated **de Broglie wavelength**, λ, which at high speeds (where you have to take into account relativistic effects (see p 98)) is approximately:

$$\lambda \simeq \frac{hc}{E}$$

4) If a beam of **high-energy electrons** is directed onto a thin film of material in front of a screen, a **diffraction pattern** will be seen on the screen.
5) As with light diffraction patterns, the first minimum appears where:

d is the diameter of the nucleus it has been scattered by.

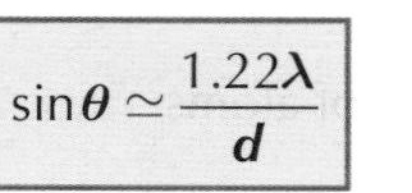

Electron beam

Electrons diffracted

Thin metal foil

θ = the angle between the straight-through position and the first minimum

Screen

first minimum

Pattern of rings seen on screen

6) Using measurements from this diffraction pattern, the **size and spacing** of the material's atomic **nuclei** can be worked out.

Example

A beam of 300 MeV electrons is fired at a piece of thin foil, and produces a diffraction pattern on a fluorescent screen. The first minimum of the diffraction pattern is at an angle of 30° from the straight-through position. Estimate the diameter of the nuclei the electrons were diffracted by.

$$E = 300 \text{ MeV} = 3.00\times10^{8}\times1.6\times10^{-19} = 4.8\times10^{-11}\text{J},\ \lambda \simeq \frac{hc}{E} = \frac{6.6\times10^{-34}\times3.0\times10^{8}}{4.8\times10^{-11}} = 4.125\times10^{-15}\text{m}$$

$$\text{So } d \simeq \frac{1.22\lambda}{\sin\theta} = \frac{1.22\times4.125\times10^{-15}}{\sin30^\circ} = 2\times1.22\times4.125\times10^{-15} = \mathbf{1.0\times10^{-14}\ m}$$

Practice Questions

Q1 Explain how alpha particle scattering shows that a nucleus is both small and positively charged.

Q2 Why are X-rays a suitable electromagnetic wave to investigate atomic sizes?

Q3 Describe how you could estimate the distance of closest approach of a scattered particle.

Exam Questions

Q1 A beam of alpha particles is directed onto a very thin gold film.

(a) Explain why the majority of alpha particles are not scattered. [2 marks]

(b) Explain how alpha particles are scattered by atomic nuclei. [3 marks]

Q2 Various particles can be used to investigate the structure of matter.

(a) Explain why particles such as electrons produce diffraction patterns. [2 marks]

(b) Electron-beam diffraction is currently the most accurate method for finding out about the atomic structure of a crystal. State why this is. [1 mark]

(c) When electrons are directed at a larger nucleus, the beam suffers less diffraction. Why does this happen? [2 marks]

Alpha scattering — It's positively repulsive...

Scattering and diffraction are the key ideas you need to understand for questions about atomic size and structure. Remember, particles like electrons have wave-like properties, so if you fire them at crystal structures they make a diffraction pattern. This lets you work out size, spacing etc...

Nuclear Radius and Density

The tiny nucleus — such a weird place, but one that you need to become ultra familiar with. Lucky you...

The Nucleus is a Very Small Part of a Whole Atom

1) By **probing atoms** using scattering and diffraction methods, we know that the **diameter of an atom** is about 0.1 nm (1×10^{-10} m) and the diameter of the smallest **nucleus** is about 2 fm (2×10^{-15} m — pronounced "femtometres").
2) So basically, **nuclei** are really, really **tiny** compared with the size of the **whole atom**.
3) To make this **easier to visualise**, try imagining a **large Ferris wheel** (which is pretty darn big) as the size of **an atom**. If you then put a **grain of rice** (which is rather small) in the centre, this would be the size of the atom's **nucleus**.
4) **Molecules** are just a number of **atoms joined together**. As a rough guide, the size of a molecule equals the number of atoms in it multiplied by the size of one atom.

The Nucleus is Made Up of Nucleons

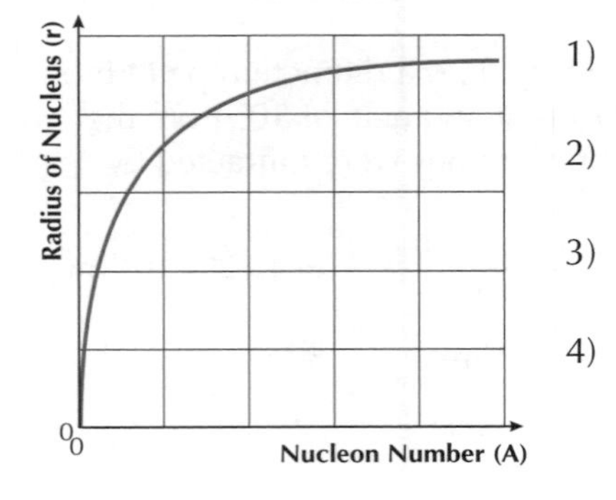

1) The **particles** that make up the nucleus (i.e. **protons** and **neutrons**) are called **nucleons**.
2) The **number of nucleons** in an atom is called the **mass** (or **nucleon**) **number, *A***.
3) As **more nucleons** are added to the nucleus, it gets **bigger**.
4) And as we all know by now, you can measure the size of a nucleus by firing particles at it (see p 32-33).

See p. 40 for more on the mass number and how this is used to represent atomic structure in standard notation.

Nuclear Radius is Proportional to the Cube Root of the Mass Number

The **nuclear radius** increases roughly as the cube root of the mass (nucleon) number.

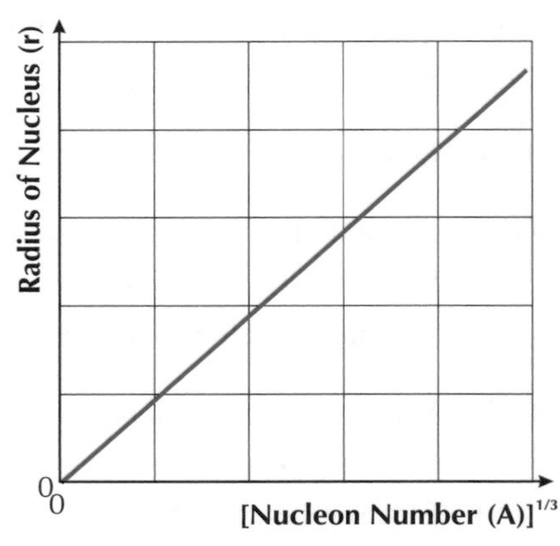

1) This **straight-line graph** shows that the **nuclear radius** (***r***) is **directly proportional** to the cube root of the **nucleon number** (***A***).
2) This relationship can be written as: $r \propto A^{1/3}$.
3) We can make this into an equation by introducing a constant, r_0, giving:

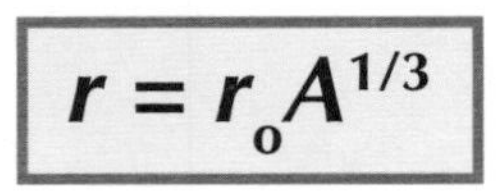

$$r = r_0 A^{1/3}$$

Where r_0 is the value of r when $A = 1$, i.e. for a proton (hydrogen nucleus). The value of r_0 is about 1.4 fm.

Example Calculate the radius of an oxygen nucleus which has 16 nucleons.

$$r = r_0 A^{1/3} = 1.4 \times 10^{-15} \times (16)^{1/3}$$
$$= 3.5 \times 10^{-15} \text{ m (or 3.5 fm)}$$

Nuclear Radius and Density

The Density of Nuclear Matter is Enormous

1) The **volume** that each nucleon (i.e. a **proton** or a **neutron**) takes up in a nucleus is about the **same**.
2) Because protons and neutrons have nearly the **same mass**, it means that all nuclei have a **similar density** (ρ).
3) But nuclear matter is **no ordinary** stuff. Its density is **enormous**. A **teaspoon** of pure nuclear matter would have a mass of about **five hundred million tonnes**. (Just to make you gasp in awe and wonder, out in space nuclear matter makes up neutron stars, which are several kilometres in diameter.)

The following examples show how **nuclear density** is pretty much the **same**, **regardless of the element**.

Example 1

Work out the density of a carbon nucleus given that its mass is 2.00×10^{-26} kg and $A = 12$.

1) The **radius** (r) of a carbon nucleus $\approx$ **3.2×10^{-15} m**
2) So, the **volume** (v) of the nucleus $= \frac{4}{3}\pi r^3$ $= \mathbf{1.37 \times 10^{-43}\ m^3}$
3) This gives the **density** (ρ) of a carbon nucleus as:

$$\rho = \frac{m}{v} = \frac{2.00 \times 10^{-26}}{1.37 \times 10^{-43}} = 1.46 \times 10^{17}\ \text{kg m}^{-3}$$

Example 2

Work out the density of a gold nucleus given that its mass is 3.27×10^{-25} kg and $A = 197$.

1) The **radius** (r) of a gold nucleus $\approx$ **8.1×10^{-15} m**
2) So, the **volume** (v) of the nucleus $= \frac{4}{3}\pi r^3$ $= \mathbf{2.23 \times 10^{-42}\ m^3}$
3) This gives the **density** (ρ) of a gold nucleus as:

$$\rho = \frac{m}{v} = \frac{3.27 \times 10^{-25}}{2.23 \times 10^{-42}} = 1.47 \times 10^{17}\ \text{kg m}^{-3}$$

Nuclear density is significantly larger than atomic density — this suggests three important facts about the structure of an atom:
a) Most of an atom's mass is in its nucleus.
b) The nucleus is small compared to the atom.
c) An atom must contain a lot of empty space.

Practice Questions

Q1 What is the approximate size of an atom?

Q2 What are nucleons?

Q3 What is the relationship between the nuclear radius and mass number?

Q4 In the formula $r = r_o A^{1/3}$, what does r_o represent?

Q5 Explain why the density of ordinary matter is much less than that of nuclear matter.

Exam Questions

Q1 The radius (r) of a nucleus with A nucleons can be calculated using the equation $r = r_o A^{1/3}$.
(a) If a carbon nucleus containing 12 nucleons has a radius of 3.2×10^{-15} m, show that $r_o = 1.4 \times 10^{-15}$ m. [2 marks]
(b) Calculate the radius of a radium nucleus containing 226 nucleons. [1 mark]
(c) Calculate the density of the radium nucleus if its mass is 3.75×10^{-25} kg. [2 marks]

Q2 A sample of pure gold has a density of 19 300 kg m^{-3}. If the density of a gold nucleus is 1.47×10^{17} kg m^{-3}, discuss what this implies about the structure of a gold atom. [4 marks]

Nuclear and particle physics — heavy stuff...

So basically the nucleus is a tiny part of the atom, but it's incredibly dense. The density doesn't change much from element to element, and the radius depends on the mass number. Learn the theory like your own backyard, but don't worry about remembering equations and values — those friendly examiners have popped them in the exam paper for you. How nice...

Radioactive Emissions

Radiation is all around us... Not quite as catchy as the original, but it is true at least...

Unstable Atoms are Radioactive

1) If an atom is **unstable**, it will **break down** to **become** more stable. Its **instability** could be caused by having **too many neutrons**, **not enough neutrons**, or just **too much energy** in the nucleus.
2) The atom **decays** by **releasing energy** and/or **particles**, until it reaches a **stable form** — this is called **radioactive decay**.
3) An individual radioactive decay is **random** — it can't be predicted.

There are Four Types of Nuclear Radiation

Learn this table.

Radiation	Symbol	Constituent	Relative Charge	Mass (u)
Alpha	α	A helium nucleus — 2 protons & 2 neutrons	+2	4
Beta-minus (Beta)	β or β⁻	Electron	-1	(negligible)
Beta-plus	β⁺	Positron	+1	(negligible)
Gamma	γ	Short-wavelength, high-frequency electromagnetic wave.	0	0

u stands for atomic mass unit — see p 42.

You met positrons at AS level.

The Different Types of Radiation have Different Penetrations

When a radioactive particle **hits** an **atom** it can **knock off electrons**, creating an **ion** — so, **radioactive emissions** are also known as **ionising radiation**.

Alpha, **beta** and **gamma** radiation can be **fired** at a **variety of objects** with **detectors** placed the **other side** to see whether they **penetrate** the object.

Skin or paper stops ALPHA

Many cm lead stops GAMMA

Thin mica

Few mm aluminium stops BETA

Radiation	Symbol	Ionising	Speed	Penetrating power	Affected by magnetic field
Alpha	α	Strongly	Slow	Absorbed by paper or a few cm of air	Yes
Beta-minus (Beta)	β or β⁻	Weakly	Fast	Absorbed by ~3 mm of aluminium	Yes
Beta-plus	β⁺	Annihilated by electron — so virtually zero range			
Gamma	γ	Very weakly	Speed of light	Absorbed by many cm of lead, or several m of concrete.	No

Alpha and Beta Particles have Different Ionising Properties

What a radioactive source can be used for often depends on its **ionising properties**.

1) **Alpha** particles are **strongly positive** — so they can **easily pull electrons** off atoms.
2) Ionising an atom **transfers** some of the **energy** from the **alpha particle** to the **atom**. The alpha particle **quickly ionises** many atoms (about 10 000 ionisations per alpha particle) and **loses** all its **energy**. This makes alpha-sources suitable for use in **smoke alarms** because they allow **current** to flow, but won't **travel very far**.
3) The **beta**-minus particle has **lower mass** and **charge** than the alpha particle, but a **higher speed**. This means it can still **knock electrons** off atoms. Each **beta** particle will ionise about 100 atoms, **losing energy** at each interaction.
4) This **lower number of interactions** means that beta radiation causes much **less damage** to body tissue.
5) Gamma radiation is even more **weakly ionising** than beta radiation, so will do even **less damage** to body tissue. This means it can be used for **diagnostic techniques** in medicine.

Radioactive Emissions

The Intensity of Gamma Radiation Obeys the Inverse Square Law

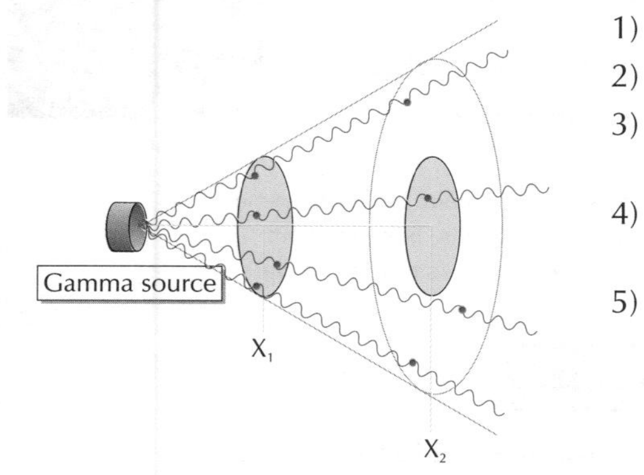

1) A **gamma source** will **emit** gamma **radiation** in **all directions**.
2) This radiation **spreads out** as you get **further away** from the source.
3) However, the amount of **radiation per unit area** (the **intensity**) will **decrease** the further you get from the source.
4) If you took a reading of **intensity**, I, at a **distance**, x, from the source you would find that it **decreases** by the **square of the distance** from the source.
5) This can be written as the equation:

$$I = \frac{kI_0}{x^2}$$

where k is a constant, and I_0 is the intensity at the source.

6) This **relationship** can be **proved** by taking **measurements of intensity** at different distances from a gamma source, using a **Geiger-Müller tube** and **counter**.
7) If the **distance** from the source is **doubled** the **intensity** is found to **fall to a quarter** — which **verifies** the inverse square law.

That's why one of the safety precautions when handling a source is to hold it at arm's length, so you lessen the amount of radiation reaching you.

We're Surrounded by Background Radiation

Put a Geiger-Müller tube **anywhere** and the counter will click — it's detecting **background radiation**.

When you take a **reading** from a radioactive source, you need to **measure** the **background radiation** separately and **subtract** it from your **measurement**.

There are many **sources** of background radiation:

1) **The air:** Radioactive **radon gas** is released from **rocks**. It emits alpha radiation. The concentration of this gas in the atmosphere varies a lot from place to place, but it's usually the largest contributor to the background radiation.
2) **The ground and buildings: All rock** contains radioactive isotopes.
3) **Cosmic radiation:** Cosmic rays are particles (mostly high-energy protons) from **space**. When they collide with particles in the upper atmosphere, they produce nuclear radiation.
4) **Living things:** All plants and animals contain **carbon**, and some of this will be radioactive **carbon-14**.
5) **Man-made radiation:** In most areas, radiation from **medical** or **industrial** sources makes up a tiny, tiny fraction of the background radiation.

Practice Questions

Q1 What makes an atom radioactive?

Q2 Name three types of nuclear radiation and give three properties of each.

Q3 Give three sources of background radiation.

Exam Questions

Q1 Briefly describe an absorption experiment to distinguish between alpha, beta and gamma radiation. You may wish to include a sketch in your answer. [4 marks]

Q2 The count rate detected by a G-M tube, 10 cm from a gamma source, is 240 counts per second. What would you expect the count rate to be at 40 cm from the source? [3 marks]

Radioactive emissions — as easy as α, β, γ...

You need to learn the different types of radiation and their properties. Remember that alpha particles are by far the most ionising and so cause more damage if they get inside your body than the same dose of any other radiation — which is one reason we don't use alpha sources as medical tracers. Learn this all really well, then go and have a brew and a bickie...

Exponential Law of Decay

Oooh look — some maths. Good.

Every Isotope Decays at a Different Rate

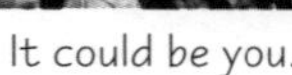
It could be you.

1) **Radioactive decay** is completely **random**. You **can't predict which** atom will decay **when**.
2) Although you can't predict the decay of an **individual atom**, if you take a **very large number of atoms**, their **overall behaviour** shows a **pattern**.
3) Any sample of a particular **isotope** has the **same rate of decay**, i.e. the same **proportion** of atoms will **decay** in a **given time**.

Isotopes of an element have the same number of protons, but different numbers of neutrons in their nuclei.

The Rate of Decay is Measured by the Decay Constant

The **activity** of a sample — the **number** of atoms that **decay each second** — is **proportional** to the **size of the sample**. For a **given isotope**, a sample **twice** as big would give **twice** the **number of decays** per second.

The **decay constant** (λ) measures how **quickly** an isotope will **decay** — the **bigger** the value of λ, the faster the rate of decay. Its unit is $\mathbf{s^{-1}}$.

activity = decay constant × number of atoms

Or in symbols: $A = \lambda N$

Don't get λ confused with wavelength.

Activity is measured in **becquerels** (Bq): 1 Bq = 1 decay per second (s^{-1})

You Need to Learn the Definition of Half-Life

The **half-life** ($T_{½}$) of an **isotope** is the **average time** it takes for the **number of undecayed atoms** to **halve**.

Measuring the **number of undecayed atoms** isn't the easiest job in the world. **In practice**, half-life isn't measured by counting atoms, but by measuring the **time it takes** the **activity** to **halve**.

The **longer** the **half-life** of an isotope, the **longer** it stays **radioactive**.

The Number of Undecayed Particles Decreases Exponentially

number of undecayed atoms: N, N/2, N/4, N/8, N/16

time, t: 0, $T_{½}$, $2T_{½}$, $3T_{½}$, $4T_{½}$

The half-life stays the same. It takes the same amount of time for half of the atoms to decay regardless of the number of atoms you start with.

The number of atoms approaches zero.

When you're **measuring** the **activity** and **half-life** of a **source**, you've got to **remember background radiation**. The **background radiation** needs to be **subtracted** from the **activity readings** to give the **source activity**.

How to find the half-life of an isotope

STEP 1: Read off the value of count rate, particles or activity when t = 0.

STEP 2: Go to half the original value.

STEP 3: Draw a horizontal line to the curve, then a vertical line down to the x-axis.

STEP 4: Read off the half-life where the line crosses the x-axis.

STEP 5: Check the units carefully.

STEP 6: It's always a good idea to check your answer. Repeat steps 1-4 for a quarter the original value. Divide your answer by two. That will also give you the half-life. Check that you get the same answer both ways.

You'd be **more likely** to actually meet a **count rate-time graph** or an **activity-time graph**. They're both **exactly the same shape** as the graph above, but with different **y-axes**.

Plotting the natural log (ln) of the number of radioactive atoms (or the activity) against time gives a straight-line graph (see p 102).

ln N, $\ln N_0$, gradient = $-\lambda$, t

Exponential Law of Decay

You Need to Know the **Equations** for **Half-Life** and **Decay**...

1) The number of radioactive nuclei decaying per second (**activity**) is proportional to the number of nuclei remaining.
2) The **half-life** can be **calculated** using the equation:
(where ln is the natural log)

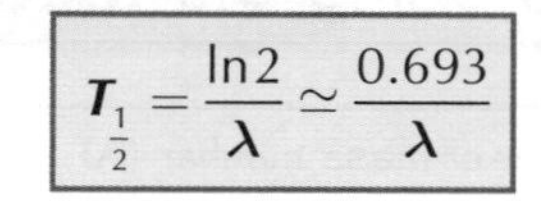

$$T_{\frac{1}{2}} = \frac{\ln 2}{\lambda} \simeq \frac{0.693}{\lambda}$$

3) The **number of radioactive atoms** remaining, N, depends on the **number originally** present, N_o. The **number remaining** can be calculated using the equation:

$$N = N_0 e^{-\lambda t}$$

Here t = time, measured in seconds.

4) As a sample decays, its activity goes down — there's an equation for that too:

$$A = A_0 e^{-\lambda t}$$

Example:
A sample of the radioactive isotope ^{13}N contains 5×10^6 atoms. The decay constant for this isotope is 1.16×10^{-3} s^{-1}.

a) What is the half-life for this isotope?

$$T_{\frac{1}{2}} = \frac{\ln 2}{1.16 \times 10^{-3}} = 598 \text{ s}$$

b) How many atoms of ^{13}N will remain after 800 seconds?

$$N = N_0 e^{-\lambda t} = 5 \times 10^6 e^{-(1.16 \times 10^{-3})(800)} = 1.98 \times 10^6 \text{ atoms}$$

Radioactive Isotopes Have **Many Uses**

Radioactive substances are extremely useful. You can use them for all sorts — to **date** organic material, diagnose **medical problems**, **sterilise** food, and in **smoke alarms**.

Radiocarbon Dating

The radioactive isotope **carbon-14** is used in **radiocarbon dating**. Living plants take in carbon dioxide from the atmosphere as part of **photosynthesis**, including the **radioactive isotope carbon-14**. When they die, the **activity** of carbon-14 in the plant starts to **fall**, with a **half-life** of around **5730 years**. Archaeological finds made from once living material (like wood) can be tested to find the **current amount** of carbon-14 in them, and date them.

Medical Diagnosis

Technetium-99m is widely used in medical tracers — **radioactive substances** that are used to show tissue or **organ function**. The tracer is **injected** into or **swallowed** by the patient and then **moves** through the **body** to the region of interest. The **radiation emitted** is **recorded** and an **image** of inside the patient produced. Technetium-99m is suitable for this use because it emits **γ-radiation**, has a **half-life of 6 hours** (long enough for data to be recorded, but short enough to limit the radiation to an acceptable level) and **decays** to a **much more stable isotope**.

Practice Questions

Q1 Define radioactive activity. What units is it measured in?

Q2 Sketch a general radioactive decay graph showing the number of undecayed particles against time.

Q3 What is meant by the term 'half-life'?

Q4 Describe how radiocarbon dating works.

Exam Questions

Q1 Explain what is meant by the random nature of radioactive decay. [1 mark]

Q2 You take a reading of 750 Bq from a pure radioactive source. The radioactive source initially contains 50 000 atoms, and background activity in your lab is measured as 50 Bq.

(a) Calculate the decay constant for your sample. [3 marks]

(b) Determine the half-life of this sample. [2 marks]

(c) Approximately how many atoms of the radioactive source will there be after 300 seconds? [2 marks]

Radioactivity is a random process — just like revision shouldn't be...

Remember the shape of that graph — whether it's count rate, activity or number of atoms plotted against time, the shape's always the same. This is all pretty straightforward mathsy-type stuff: plugging values in equations, reading off graphs, etc. Not very interesting, though. Ah well, once you get onto relativity you'll be longing for a bit of boredom.

Nuclear Decay

The stuff on these pages covers the most important facts about nuclear decay that you're just going to have to make sure you know inside out. I'd be very surprised if you didn't get a question about it in your exam...

Atomic Structure can be Represented Using Standard Notation

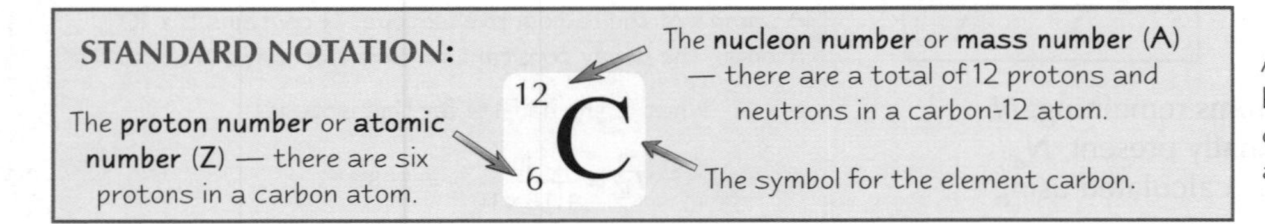

Atoms with the **same number of protons** but **different numbers of neutrons** are called **isotopes**. The following examples are all isotopes of carbon: ${}^{12}_{6}C$, ${}^{13}_{6}C$, ${}^{14}_{6}C$

Some Nuclei are More Stable than Others

The nucleus is under the **influence** of the **strong nuclear force holding** it **together** and the **electromagnetic force pushing** the **protons apart**. It's a very **delicate balance**, and it's easy for a nucleus to become **unstable**. You can get a stability graph by plotting **Z** (atomic number) against **N** (number of neutrons).

A nucleus will be **unstable** if it has:

1) **too many neutrons**
2) **too few neutrons**
3) **too many nucleons** altogether, i.e. it's **too heavy**
4) **too much energy**

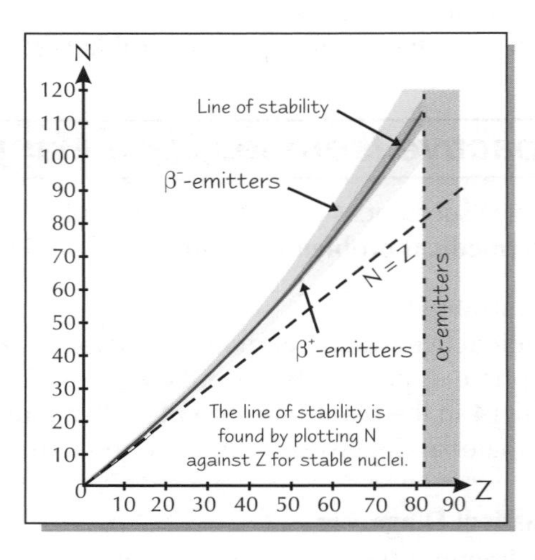

α Emission Happens in Heavy Nuclei

When an alpha particle is **emitted**:

The **proton number decreases** by **two**, and the **nucleon number decreases** by **four**.

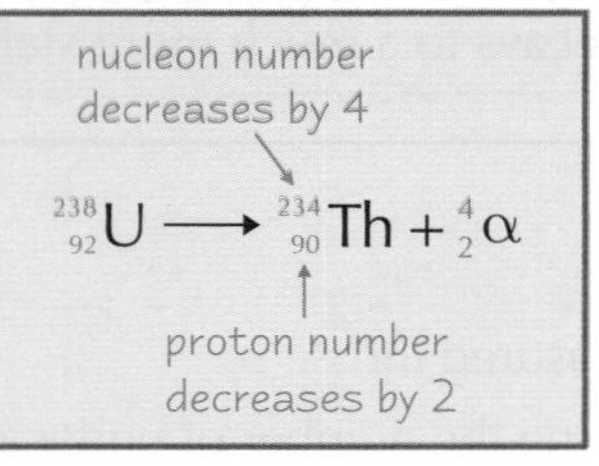

1) **Alpha emission** only happens in **very heavy** atoms (with more than 82 protons), like **uranium** and **radium**.
2) The **nuclei** of these atoms are **too massive** to be stable.

β⁻ Emission Happens in Neutron Rich Nuclei

1) **Beta-minus** (usually just called beta) decay is the emission of an **electron** from the **nucleus** along with an **antineutrino**.
2) Beta decay happens in isotopes that are **"neutron rich"** (i.e. have many more **neutrons** than **protons** in their nucleus).
3) When a nucleus ejects a beta particle, one of the **neutrons** in the nucleus is **changed** into a **proton**.

When a **beta-minus** particle is **emitted**:

The **proton number increases** by **one**, and the **nucleon number stays the same**.

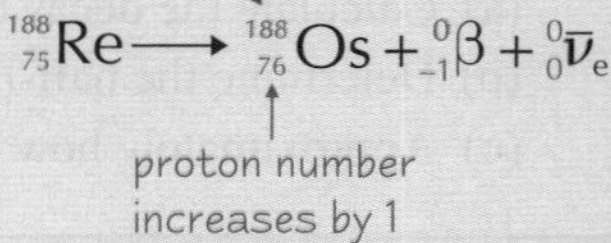

In **beta-plus emission**, a **proton** gets **changed** into a **neutron**.
The **proton number decreases** by **one**, and the **nucleon number stays the same**.

Nuclear Decay

γ Radiation is Emitted from Nuclei with Too Much Energy

1) **After alpha** or **beta** decay, the **nucleus** often has **excess energy** — it's **excited**. This energy is **lost** by emitting a **gamma ray**.
2) **Another way** that gamma radiation is produced is when a nucleus **captures** one of its own orbiting **electrons**.

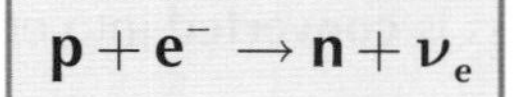

$$p + e^- \rightarrow n + \nu_e$$

3) **Electron capture** causes a **proton** to **change** into a **neutron**. This makes the **nucleus unstable** and it **emits** gamma radiation.

The artificial isotope technetium-99^m is formed in an excited state from the decay of another element. It is used as a tracer in medical imaging (see p 39).

During **gamma emission**, there is **no change** to the nuclear **constituents** — the nucleus just **loses excess energy**.

There are Conservation Rules in Nuclear Reactions

In every nuclear reaction **energy**, **momentum**, **proton number / charge** and **nucleon number** must be conserved.

238 = 234 + 4 — nucleon numbers balance

$${}^{238}_{92}U \rightarrow {}^{234}_{90}Th + {}^{4}_{2}\alpha$$

92 = 90 + 2 — proton numbers balance

Mass is Not Conserved

1) The **mass** of the **alpha particle** is less than the **individual masses** of **two protons** and **two neutrons**. The difference is called the **mass defect**.
2) Mass **doesn't** have to be **conserved** because of **Einstein's equation**:

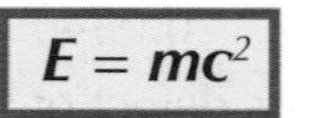

$$E = mc^2$$

3) This says that **mass and energy** are **equivalent**.
The **energy released** when the nucleons **bonded together** accounts for the missing mass — so the **energy released** is the same as the **mass defect** × c^2.

Practice Questions

Q1 What makes a nucleus unstable? Describe the changes that happen in the nucleus during alpha, beta and gamma decay.

Q2 Explain the circumstances in which gamma radiation may be emitted.

Q3 Define the mass defect.

Exam Questions

Q1 (a) Radium-226 undergoes alpha decay to radon. Complete the balanced nuclear equation for this reaction.

$${}^{226}_{88}Ra \rightarrow \quad Rn +$$ [3 marks]

(b) Potassium-40 ($Z = 19$, $A = 40$) undergoes beta decay to calcium.
Write a balanced nuclear equation for this reaction. [3 marks]

Q2 Calculate the energy released during the formation of an alpha particle, given that the total mass of two protons and two neutrons is 6.695×10^{-27} kg, the mass of an alpha particle is 6.645×10^{-27} kg and the speed of light, c, is 3.00×10^8 ms^{-1}. [3 marks]

Nuclear decay — it can be enough to make you unstable...

$E = mc^2$ is an important equation that says mass and energy are equivalent. Remember it well, 'cos you're going to come across it a lot in questions about mass defect and the energy released in nuclear reactions over the next few pages...

Binding Energy

Turn off the radio and close the door, 'cos you're going to need to concentrate hard on this stuff about binding energy...

The Mass Defect is Equivalent to the Binding Energy

1) The **mass** of a **nucleus** is **less than** the mass of its **constituent parts** — the difference is called the **mass defect** (see p. 41).
2) Einstein's equation, $E = mc^2$, says that mass and energy are **equivalent**.
3) So, as nucleons join together, the total mass **decreases** — this '**lost**' mass is **converted** into energy and **released**.
4) The amount of **energy released** is **equivalent** to the **mass defect**.
5) If you **pulled** the nucleus completely **apart**, the **energy** you'd have to use to do it would be the **same** as the energy **released** when the nucleus formed.

> The energy needed to **separate** all of the nucleons in a nucleus is called the **binding energy** (measured in **MeV**), and it is **equivalent** to the **mass defect**.

Example Calculate the binding energy of the nucleus of a lithium atom, $^{6}_{3}\text{Li}$, given that its mass defect is 0.0343 u.

1) Convert the mass defect into kg.

   Mass defect $= 0.0343 \times 1.66 \times 10^{-27} = 5.70 \times 10^{-29}$ kg

2) Use $E = mc^2$ to calculate the binding energy.

   $E = 5.70 \times 10^{-29} \times (3 \times 10^{8})^2 = 5.13 \times 10^{-12}$ J = 32 MeV

Atomic mass is usually given in atomic mass units (u), where $1\text{ u} = 1.66 \times 10^{-27}$ kg.

$1\text{ MeV} = 1.6 \times 10^{-13}$ J

6) The **binding energy per unit of mass defect** can be calculated (using the example above):

$$\frac{\text{binding energy}}{\text{mass defect}} = \frac{32\text{ MeV}}{0.0343\text{ u}} \approx 931.3\text{ MeV u}^{-1}$$

7) This means that a mass defect of **1 u** is equivalent to about **931.3 MeV** of binding energy.

Captain Skip didn't believe in ghosts, marmalade and that things could be bound without rope.

The Binding Energy Per Nucleon is at a Maximum around N = 50

A useful way of **comparing** the binding energies of different nuclei is to look at the **binding energy per nucleon**.

> Binding energy per nucleon (in MeV) $= \dfrac{\text{Binding energy (B)}}{\text{Nucleon number (A)}}$

So, the binding energy per nucleon for $^{6}_{3}\text{Li}$ (in the example above) is $32 \div 6 = 5.3$ MeV.

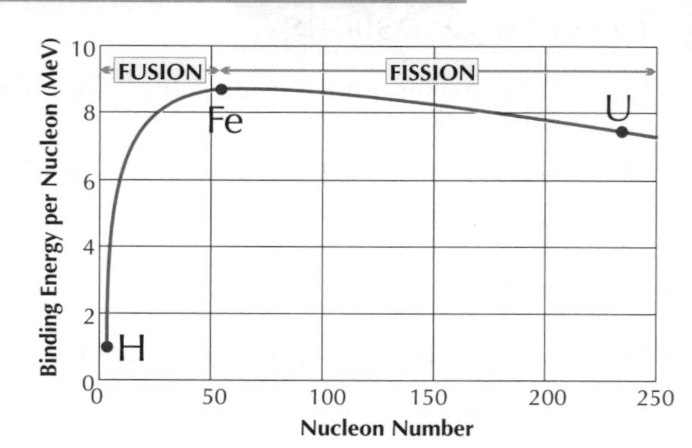

1) A **graph** of **binding energy per nucleon** against **nucleon number**, for all elements, shows a **curve**.

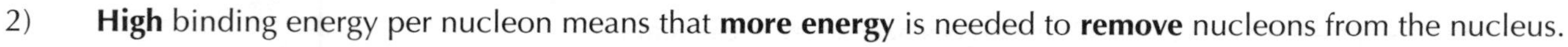

2) **High** binding energy per nucleon means that **more energy** is needed to **remove** nucleons from the nucleus.
3) In other words the **most stable** nuclei occur around the **maximum point** on the graph — which is at **nucleon number 56** (i.e. **iron**, Fe).
4) **Combining small nuclei** is called nuclear **fusion** (see p. 45) — this **increases** the **binding energy per nucleon** dramatically, which means a lot of **energy is released** during nuclear fusion.
5) **Fission** is when **large nuclei** are **split in two** (see p. 44) — the **nucleon numbers** of the two **new nuclei** are **smaller** than the original nucleus, which means there is an **increase** in the binding energy per nucleon. So, energy is also **released** during nuclear fission (but not as much energy per nucleon as in nuclear fusion).

Binding Energy

The Change in Binding Energy Gives the Energy Released...

The **binding energy per nucleon graph** can be used to **estimate** the **energy released** from nuclear reactions.

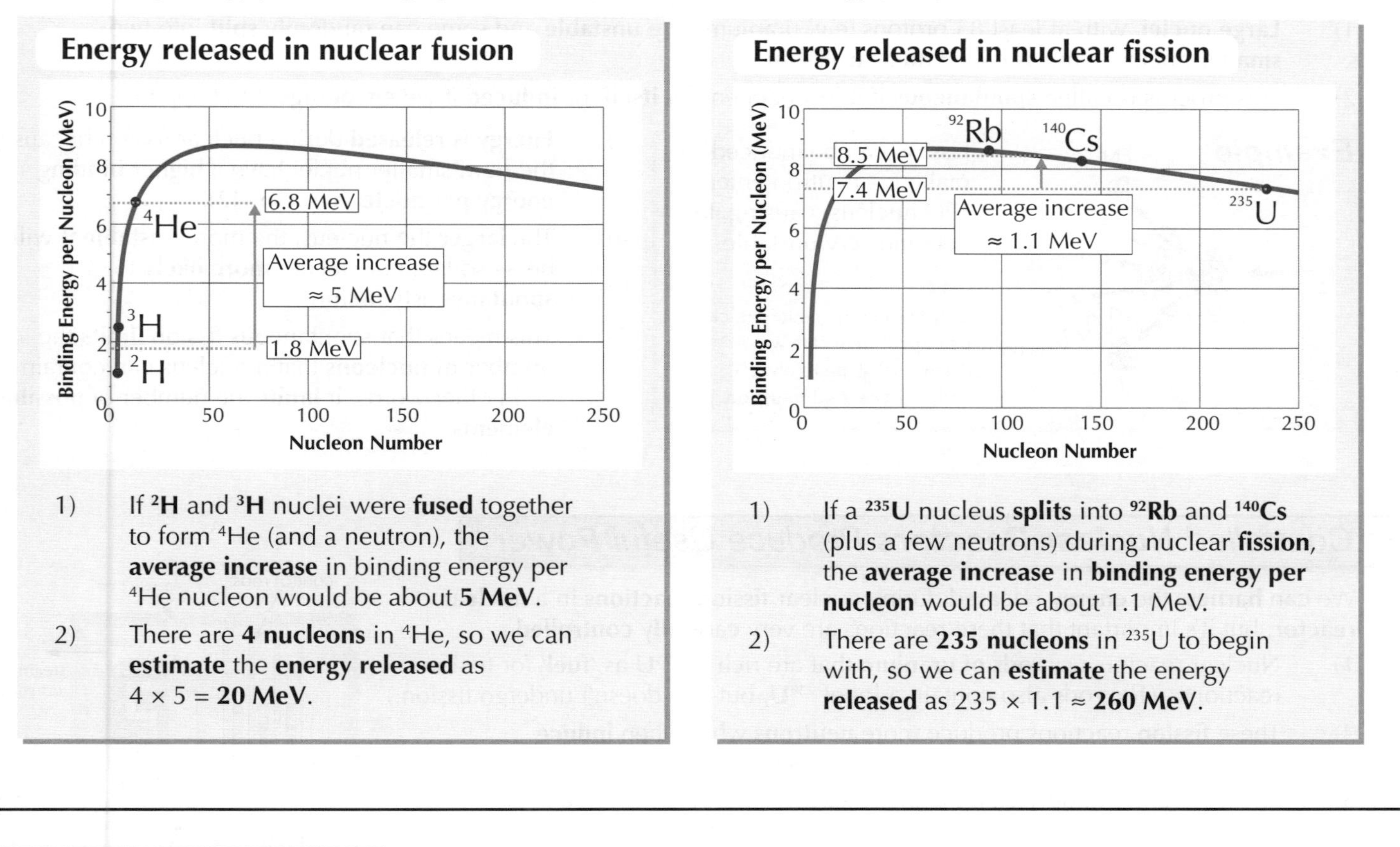

Energy released in nuclear fusion

1) If ^{2}**H** and ^{3}**H** nuclei were **fused** together to form ^{4}He (and a neutron), the **average increase** in binding energy per ^{4}He nucleon would be about **5 MeV**.
2) There are **4 nucleons** in ^{4}He, so we can **estimate** the **energy released** as $4 \times 5 =$ **20 MeV**.

Energy released in nuclear fission

1) If a ^{235}**U** nucleus **splits** into ^{92}**Rb** and ^{140}**Cs** (plus a few neutrons) during nuclear **fission**, the **average increase** in **binding energy per nucleon** would be about 1.1 MeV.
2) There are **235 nucleons** in ^{235}U to begin with, so we can **estimate** the energy **released** as $235 \times 1.1 \approx$ **260 MeV**.

Practice Questions

Q1 What is the binding energy of a nucleus?

Q2 How can we calculate the binding energy for a particular nucleus?

Q3 What is the binding energy per nucleon?

Q4 Which element has the highest value of binding energy per nucleon?

Q5 Do nuclear fusion or fission reactions release more energy per nucleon?

Exam Questions

Q1 The mass of a $^{14}_{6}$**C** nucleus is 13.999948 u. The mass of a proton is 1.007276 u, and a neutron is 1.008665 u.

(a) Calculate the mass defect of a $^{14}_{6}$**C** nucleus (given that 1 u = 1.66×10^{-27} kg). [3 marks]

(b) Use $E = mc^2$ to calculate the binding energy of the nucleus in MeV (given that $c = 3 \times 10^8$ ms^{-1} and 1 MeV = 1.6×10^{-13} J). [2 marks]

Q2 The following equation represents a nuclear reaction that takes place in the Sun:

$${}^1_1\text{p} + {}^1_1\text{p} \rightarrow {}^2_1\text{H} + {}^{\ 0}_{+1}\beta + \text{energy released}$$

where p is a proton and β is a positron (opposite of an electron)

(a) State the type of nuclear reaction shown. [1 mark]

(b) Given that the binding energy per nucleon for a proton is 0 MeV and for a ^{2}H nucleus it is approximately 0.86 MeV, estimate the energy released by this reaction. [2 marks]

A mass defect of 1 u is equivalent to a binding energy of 931.3 MeV...

Remember this useful little fact, and it'll save loads of time in the exam — because you won't have to fiddle around with converting atomic mass from u → kg and binding energy from J → MeV. What more could you possibly want...

Nuclear Fission and Fusion

What did the nuclear scientist have for his tea? Fission chips... hohoho.

Fission Means Splitting Up into Smaller Parts

1) **Large nuclei**, with at least 83 protons (e.g. uranium), are **unstable** and some can randomly **split** into two **smaller** nuclei — this is called **nuclear fission**.
2) This process is called **spontaneous** if it just happens **by itself**, or **induced** if we **encourage** it to happen.

Example

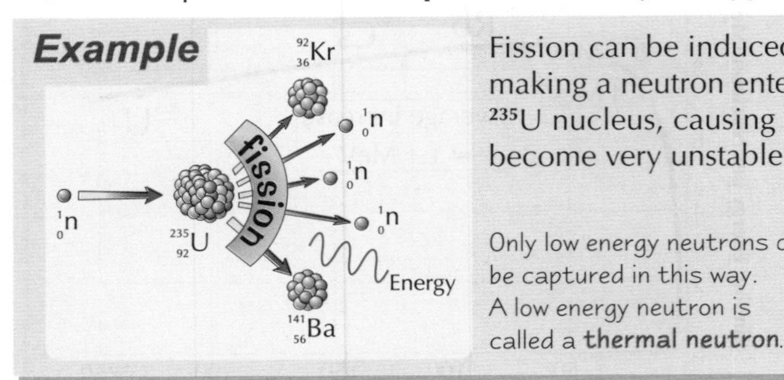

Fission can be induced by making a neutron enter a ^{235}U nucleus, causing it to become very unstable.

Only low energy neutrons can be captured in this way. A low energy neutron is called a **thermal neutron**.

3) **Energy is released** during nuclear fission because the new, smaller nuclei have a **higher binding energy per nucleon** (see p. 42).
4) The **larger** the nucleus, the more **unstable** it will be — so large nuclei are **more likely** to **spontaneously fission**.
5) This means that spontaneous fission **limits** the **number of nucleons** that a nucleus can contain — in other words, it **limits** the number of **possible elements**.

Controlled Nuclear Reactors Produce Useful Power

We can **harness** the **energy** released during nuclear **fission reactions** in a **nuclear reactor**, but it's important that these reactions are very **carefully controlled**.

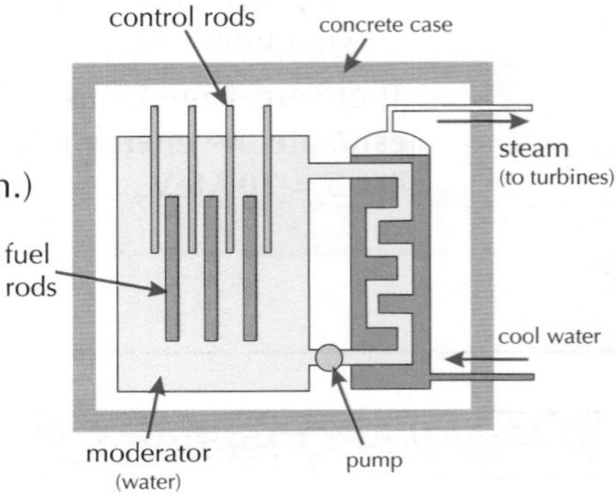

1) Nuclear reactors use **rods of uranium** that are rich in **^{235}U** as **'fuel'** for fission reactions. (The rods also contain a lot of **^{238}U**, but that doesn't undergo fission.)
2) These **fission** reactions produce more **neutrons** which then **induce** other nuclei to fission — this is called a **chain reaction**.
3) The **neutrons** will only cause a chain reaction if they are **slowed down**, which allows them to be **captured** by the uranium nuclei — these slowed down neutrons are called **thermal neutrons**.
4) ^{235}U **fuel rods** need to be placed in a **moderator** (for example, **water**) to **slow down** and/or absorb **neutrons**. You need to choose a moderator that will slow down some neutrons enough so they can cause **further fission**, keeping the reaction going at a steady rate. Choosing a moderator that absorbs **more neutrons the higher the temperature** will **decrease** the chance of **meltdown** if the reactor **overheats** — as it will naturally **slow down** the reaction.
5) You want the chain reaction to continue on its own at a **steady rate**, where **one** fission follows another. The amount of 'fuel' you need to do this is called the **critical mass** — any less than the critical mass (**sub-critical mass**) and the reaction will just peter out. Nuclear reactors use a **supercritical** mass of fuel (where several new fissions normally follow each fission) and **control the rate of fission** using **control rods**.
6) Control rods control the **chain reaction** by **limiting** the number of **neutrons** in the reactor. These **absorb neutrons** so that the **rate of fission** is controlled. **Control rods** are made up of a material that **absorbs neutrons** (e.g. boron), and they can be inserted by varying amounts to control the reaction rate.
   In an **emergency**, the reactor will be **shut down** automatically by the **release of the control rods** into the reactor, which will stop the reaction as quickly as possible.
7) **Coolant** is sent around the reactor to **remove heat** produced in the fission — often the coolant is the **same water** that is being used in the reactor as a **moderator**. The **heat** from the reactor can then be used to make **steam** for powering **electricity-generating turbines**.
8) The nuclear reactor is surrounded by a thick **concrete case**, which acts as **shielding**. This prevents **radiation escaping** and reaching the people working in the power station.

If the chain reaction in a nuclear reactor is **left to continue unchecked**, large amounts of **energy** are **released** in a very **short time**.
Many new fissions will follow each fission, causing a **runaway reaction** which could lead to an **explosion**.
This is what happens in a **fission (atomic) bomb**.

Nuclear Fission and Fusion

Waste Products of Fission Must be Disposed of Carefully

1) The **waste products** of **nuclear fission** usually have a **larger proportion of neutrons** than nuclei of a similar atomic number — this makes them **unstable** and **radioactive**.
2) The products can be used for **practical applications** such as **tracers** in medical diagnosis (see p39).
3) However, they may be **highly radioactive** and so their **handling** and **disposal** needs **great care**.
4) When material is removed from the reactor, it is initially **very hot**, so is placed in **cooling ponds** until the **temperature falls** to a safe level.
5) The radioactive waste is then **stored** underground in **sealed containers** until its **activity has fallen** sufficiently.

Fusion Means Joining Nuclei Together

1) **Two light nuclei** can **combine** to create a larger nucleus — this is called **nuclear fusion**.
2) A lot of **energy** is released during nuclear fusion because the new, heavier nuclei have a **much higher binding energy per nucleon** (see p. 42).

Example

In the Sun, **hydrogen nuclei** fuse in a series of reactions to form **helium**.

$${}^{2}_{1}H + {}^{1}_{1}H \rightarrow {}^{3}_{2}He + energy$$

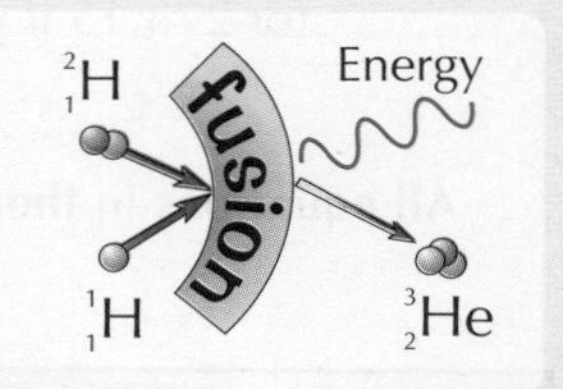

Nuclei Need Lots of Energy to Fuse

1) All nuclei are **positively charged** — so there will be an **electrostatic** (or Coulomb) **force** of **repulsion** between them.
2) Nuclei can only **fuse** if they **overcome** this electrostatic force and get **close** enough for the attractive force of the **strong interaction** to hold them both together.
3) About **1 MeV** of kinetic energy is **needed** to make nuclei fuse together — and that's **a lot of energy**.

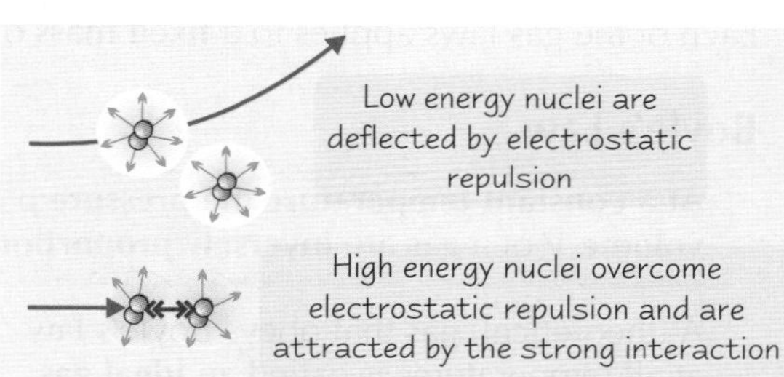

Practice Questions

Q1 What is spontaneous fission?

Q2 How can fission be induced in ^{235}U?

Q3 Why must the waste products of nuclear fission be disposed of very carefully?

Q4 Why is a lot of energy required for nuclear fusion to occur?

Exam Questions

Q1 Nuclear reactors use carefully controlled chain reactions to produce energy.

(a) Explain what is meant by the expression 'chain reaction' in terms of nuclear fission. [2 marks]

(b) Describe and explain one feature of a nuclear reactor whose role is to control the rate of fission. Include an example of a suitable material for the feature you have chosen. [3 marks]

(c) Explain what happens in a nuclear reactor during an emergency shut-down. [2 marks]

Q2 Discuss two advantages and two disadvantages of using nuclear fission to produce electricity. [4 marks]

If anyone asks, I've gone fission... that joke never gets old...

So, controlled nuclear fission reactions can provide a shedload of energy to generate electricity. There are pros and cons to using fission reactors... But then, you already knew that — now you need to learn all the grisly details.

Ideal Gases

*Aaahh... great... another one of those 'our equation doesn't work properly with **real gases**, so we'll invent an **ideal** gas that it **does work** for and they'll think we're dead clever' situations. Hmm. Physicists, eh...*

There's an Absolute Scale of Temperature

There is a **lowest possible temperature** called **absolute zero***. Absolute zero is given a value of **zero kelvin**, written **0 K**, on the absolute temperature scale.

At **0 K** all particles have the **minimum** possible **kinetic energy** — everything pretty much stops — at higher temperatures, particles have more energy. In fact, with the **Kelvin scale**, a particle's **energy** is **proportional** to its **temperature** (see page 50).

Equivalent temperatures

373 K — 100 °C

273 K — 0 °C

0 K — –273 °C

1) The Kelvin scale is named after Lord Kelvin who first suggested it.
2) A change of **1 K** equals a change of **1 °C**.
3) To change from degrees Celsius into kelvin you **add 273** (or 273.15 if you need to be really precise).

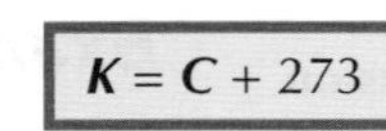

$K = C + 273$

All equations in **thermal physics** use temperatures measured in kelvin.

It's true. –273.15 °C is the lowest temperature theoretically possible. Weird, huh. You'd kinda think there wouldn't be a minimum, but there is.

There are Three Gas Laws

The three gas laws were each worked out **independently** by **careful experiment**. Each of the gas laws applies to a **fixed mass** of gas.

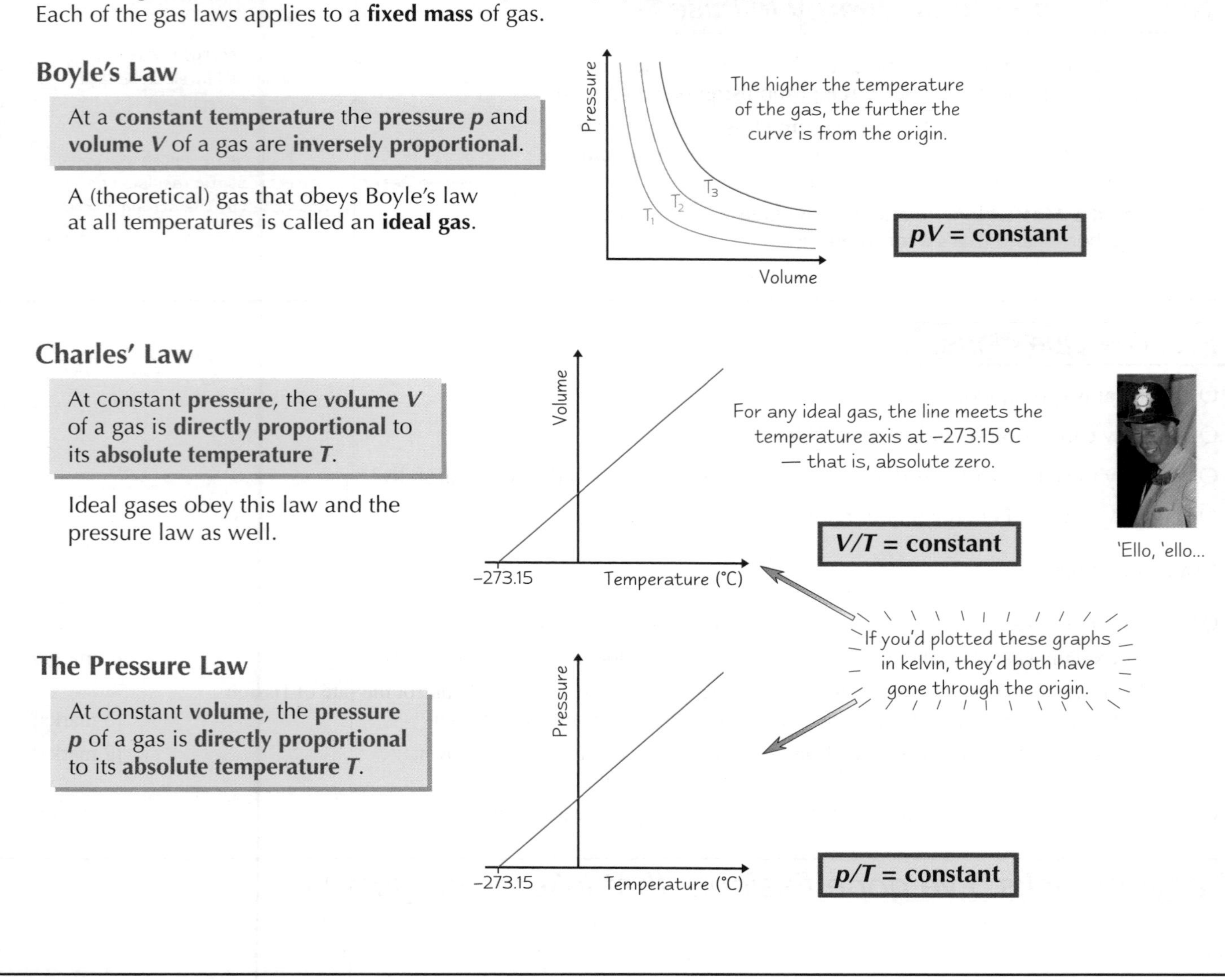

Boyle's Law

At a **constant temperature** the **pressure** p and **volume** V of a gas are **inversely proportional**.

A (theoretical) gas that obeys Boyle's law at all temperatures is called an **ideal gas**.

$pV = \text{constant}$

Charles' Law

At constant **pressure**, the **volume** V of a gas is **directly proportional** to its **absolute temperature** T.

Ideal gases obey this law and the pressure law as well.

$V/T = \text{constant}$

The Pressure Law

At constant **volume**, the **pressure** p of a gas is **directly proportional** to its **absolute temperature** T.

$p/T = \text{constant}$

Ideal Gases

If you Combine All Three you get the Ideal Gas Equation

Combining all three gas laws gives the equation: $\frac{pV}{T} =$ **constant**

1) The constant in the equation depends on the amount of gas used. ← (Pretty obvious... if you have more gas it takes up more space.)
   The amount of **gas** can be **measured** in **moles**, ***n***.
2) The constant then becomes ***nR***, where ***R*** is called the **molar gas constant**.
   Its value is 8.31 $\text{J mol}^{-1}\text{K}^{-1}$.
3) Plugging this into the equation gives: $\frac{pV}{T} = nR$ or rearranging, $pV = nRT$ — ***the ideal gas equation***

This equation works well (i.e., a real gas approximates to an ideal gas) for gases at **low pressure** and fairly **high temperatures**.

Boltzmann's Constant k is like a Gas Constant for One Particle of Gas

One mole of any **gas** contains the same number of particles.
This number is called **Avogadro's constant** and has the symbol N_A. The value of N_A is $\mathbf{6.02 \times 10^{23}}$ **particles per mole**.

1) The **number of particles** in a **mass of gas** is given by the **number of moles**, ***n***, multiplied by **Avogadro's constant**.
   So the number of particles, $N = nN_A$.
2) **Boltzmann's constant**, ***k***, is equivalent to R/N_A — you can think of Boltzmann's constant as the **gas constant** for **one particle of gas**, while ***R*** is the gas constant for **one mole of gas**.
3) The value of Boltzmann's constant is $\mathbf{1.38 \times 10^{-23}}$ $\mathbf{JK^{-1}}$.
4) If you combine $N = nN_A$ and $k = R/N_A$ you'll see that $Nk = nR$ — which can be substituted into the ideal gas equation: → $pV = NkT$ — ***the equation of state***

   The equation $pV = NkT$ is called the equation of state of an ideal gas.

Practice Questions

Q1 State Boyle's law, Charles' law and the pressure law.

Q2 What is the ideal gas equation?

Q3 The pressure of a gas is 100 000 Pa and its temperature is 27 °C. The gas is heated — its volume stays fixed but the pressure rises to 150 000 Pa. Show that its new temperature is 177 °C.

Q4 What is the equation of state of an ideal gas?

Exam Questions

Q1 The mass of one mole of nitrogen gas is 0.028 kg. R = 8.31 $\text{J mol}^{-1}\text{K}^{-1}$.

(a) A flask contains 0.014 kg of nitrogen gas. State the number of:

i) moles of nitrogen gas in the flask. [1 mark]

ii) nitrogen molecules in the flask. [1 mark]

(b) The flask has a volume of 0.01 m^3 and is at a temperature of 27 °C. Calculate the pressure inside it. [2 marks]

(c) Explain what would happen to the pressure if the number of molecules of nitrogen in the flask was halved. [2 marks]

Q2 A large helium balloon has a volume of 10 m^3 at ground level.
The temperature of the gas in the balloon is 293 K and the pressure is 1×10^5 Pa.
The balloon is released and rises to a height where its volume becomes 25 m^3 and its temperature is 260 K.
Calculate the pressure inside the balloon at its new height. [3 marks]

Ideal revision equation — marks = (pages read × questions answered)^2...

All this might sound a bit theoretical, but most gases you'll meet in the everyday world come fairly close to being 'ideal'. They only stop obeying these laws when the pressure's too high or they're getting close to their boiling point.

The Pressure of an Ideal Gas

Kinetic theory *tries to* ***explain*** *the* ***gas laws****. It basically models a gas as a series of hard balls that obey Newton's laws.*

You Need to be Able to ***Derive*** *the* ***Pressure*** *of an* ***Ideal Gas***

Start by **Deriving** the **Pressure** on **One Wall** of a Box — in the x direction

Imagine a cubic box with sides of length l containing N particles each of mass m.

This isn't an easy page. Work through it properly and make sure you understand it.

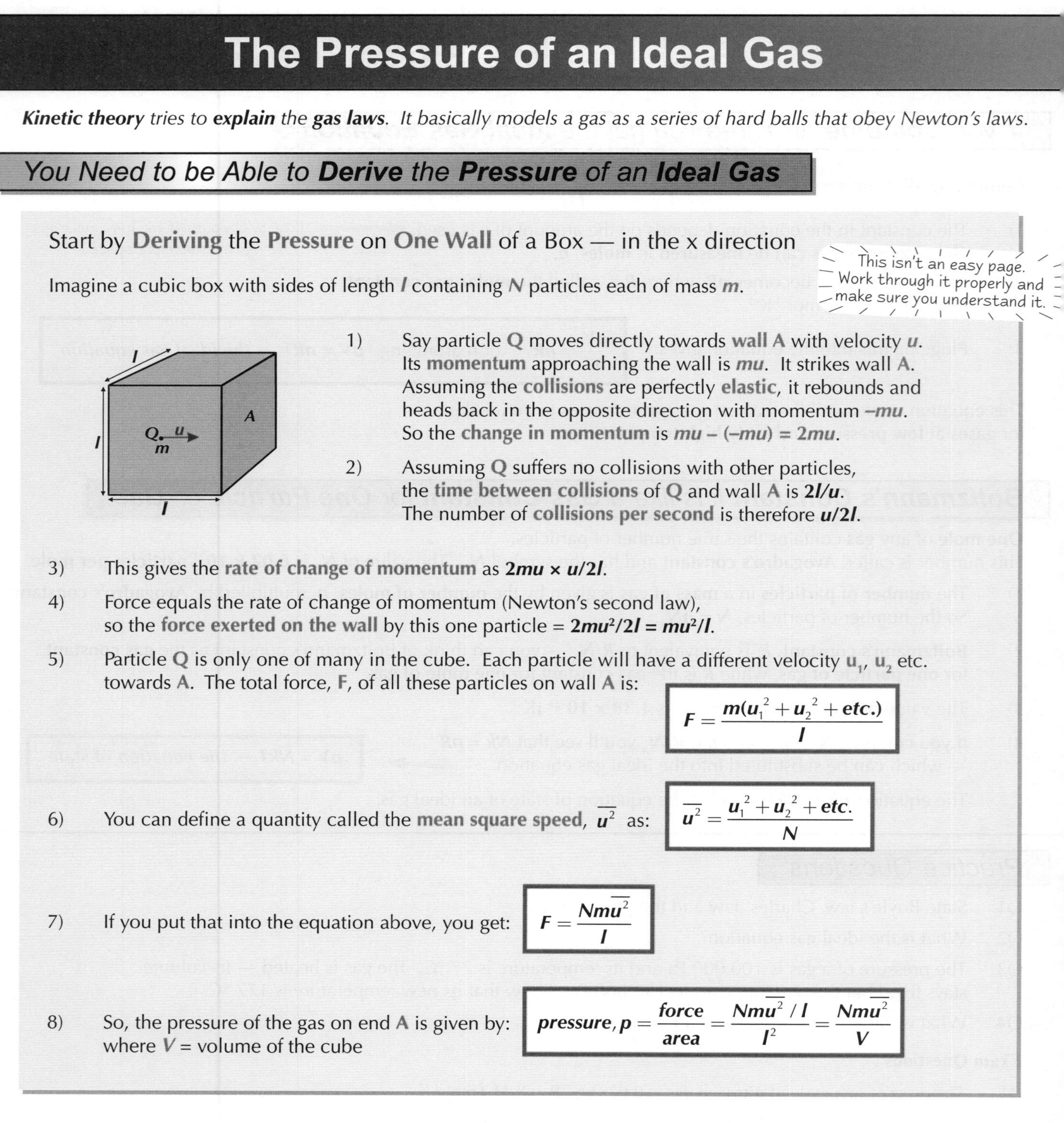

1) Say particle **Q** moves directly towards **wall A** with velocity u. Its **momentum** approaching the wall is mu. It strikes wall A. Assuming the **collisions** are perfectly **elastic**, it rebounds and heads back in the opposite direction with momentum $-mu$. So the **change in momentum** is $mu - (-mu) = 2mu$.
2) Assuming Q suffers no collisions with other particles, the **time between collisions** of Q and wall A is $2l/u$. The number of **collisions per second** is therefore $u/2l$.
3) This gives the **rate of change of momentum** as $2mu \times u/2l$.
4) Force equals the rate of change of momentum (Newton's second law), so the **force exerted on the wall** by this one particle $= 2mu^2/2l = mu^2/l$.
5) Particle Q is only one of many in the cube. Each particle will have a different velocity u_1, u_2 etc. towards A. The total force, F, of all these particles on wall A is: $$F = \frac{m(u_1^2 + u_2^2 + \text{etc.})}{l}$$
6) You can define a quantity called the **mean square speed**, $\overline{u^2}$ as: $$\overline{u^2} = \frac{u_1^2 + u_2^2 + \text{etc.}}{N}$$
7) If you put that into the equation above, you get: $$F = \frac{Nm\overline{u^2}}{l}$$
8) So, the pressure of the gas on end A is given by: $$\text{pressure}, p = \frac{\text{force}}{\text{area}} = \frac{Nm\overline{u^2}/l}{l^2} = \frac{Nm\overline{u^2}}{V}$$ where V = volume of the cube

...Then for the ***General Equation*** *you need to think about* ***All 3 Directions*** *— x, y and z*

A gas particle can move in **three dimensions** (i.e. the x, y and z directions).

1) You can calculate its **speed**, c, from Pythagoras' theorem:

   $c^2 = u^2 + v^2 + w^2$ where u, v and w are the components of the particle's velocity in the x, y and z directions.
2) If you treat all N particles in the same way, this gives an **overall** mean square speed of: $\overline{c^2} = \overline{u^2} + \overline{v^2} + \overline{w^2}$
3) Since the particles move **randomly**: $\overline{u^2} = \overline{v^2} = \overline{w^2}$ and so $\overline{c^2} = 3\overline{u^2}$
4) You can substitute this into the equation for pressure that you derived above to give: $$pV = \frac{1}{3}Nm\overline{c^2}$$

The Pressure of an Ideal Gas

A Useful Quantity is the Root Mean Square Speed $\sqrt{\overline{c^2}}$

As you saw on the previous page, it often helps to think about the motion of a **typical particle** in kinetic theory.

1) $\overline{c^2}$ is the **mean square speed** and has **units m^2s^{-2}**.
2) $\overline{c^2}$ is the **square** of the **speed** of an **average particle**, so the square root of it gives you the typical speed.
3) This is called the **root mean square speed** or, usually, the **r.m.s. speed**.
   The **unit** is the same as any speed — **ms^{-1}**.

$$\text{r.m.s. speed} = \sqrt{\text{mean square speed}} = \sqrt{\overline{c^2}}$$

Lots of Simplifying Assumptions are Used in Kinetic Theory

In **kinetic theory**, physicists picture gas particles moving at **high speed** in **random directions**.
To get **equations** like the one you just derived though, some **simplifying assumptions** are needed:

1) The gas contains a **large number of particles**.
2) The particles **move rapidly** and **randomly**.
3) The motion of the particles follows **Newton's laws**.
4) **Collisions** between particles themselves or at the walls of a container are **perfectly elastic**.
5) There are **no attractive forces** between particles.
6) Any **forces** that act during collisions are **instantaneous**.
7) Particles have a **negligible volume** compared with the volume of the container.

A **gas obeying** these **assumptions** is called an **ideal** gas. Real gases behave like ideal gases as long as the **pressure isn't too big** and the **temperature** is **reasonably high** (compared with their boiling point), so they're useful assumptions.

Practice Questions

Q1 What is the change in momentum when a gas particle hits a wall of its container head-on?

Q2 What is the force exerted on the wall by this one particle? What is the total force exerted on the wall?

Q3 What is the pressure exerted on this wall? What is the total pressure on the container?

Q4 What is 'root mean square speed'? How would you find it?

Q5 What are the seven assumptions made about ideal gas behaviour?

Exam Question

Q1 Some helium gas is contained in a flask of volume 7×10^{-5} m³. Each helium atom has a mass of 6.6×10^{-27} kg, and there are 2×10^{22} atoms present. The pressure of the gas is 1×10^{5} Pa.

(a) Calculate the mean square speed of the atoms. [2 marks]

(b) State the r.m.s. speed of a typical atom. [1 mark]

(c) If the absolute temperature of the gas is doubled, what will the r.m.s. speed of an atom become? [2 marks]

Help — these pages are de-riving me crazy...

Make sure you know the derivation inside out and back to front — it's not easy, so you might want to go through it a few times, but it is worth it. Oh, and remember — mean square speed is the average of the squared speeds — i.e. square all the speeds, then find the average. Don't make the mistake of finding the average speed first and then squaring. No, no, no...

Energy and Temperature

The energy of a particle depends on its temperature on the ***thermodynamic scale*** *(that's Kelvin to you and me).*

The *Speed Distribution* of *Gas Particles* Depends on *Temperature*

The **particles** in a **gas don't** all **travel** at the **same speed**.
Some particles will be moving fast but others much more slowly.
Most will travel around the average speed. The shape of the **speed distribution** depends on the **temperature** of the gas.

As the temperature of the gas increases:

1) the **average** particle speed increases.
2) the **maximum** particle speed increases.
3) the distribution curve becomes more **spread out**.

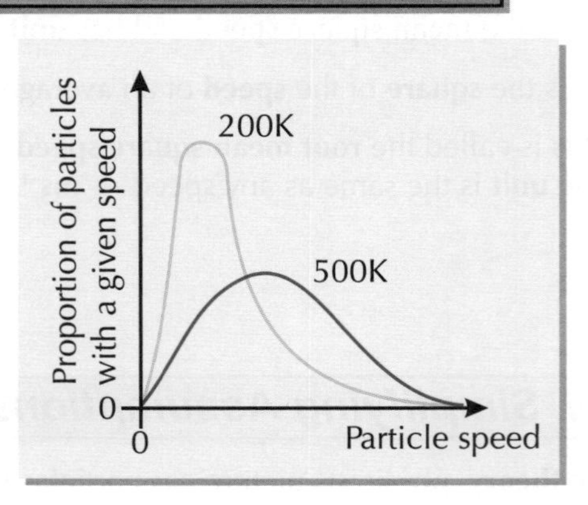

Energy Changes Happen Between Particles

The particles of a gas **collide** with each other **all the time**. Some of these collisions will be '**head-on**' (particles moving in **opposite directions**) while others will be '**shunts from behind**' (particles moving in the **same direction**).

1) As a result of the collisions, **energy** will be **transferred** between particles.
2) Some particles will **gain speed** in a collision and others will **slow down**.
3) **Between collisions**, the particles will travel at **constant speed**.
4) The energy of individual particles changes at each collision, but the **total energy** of the **system** doesn't change.
5) So, the **average** speed of the particles will stay the same provided the **temperature** of the gas **stays the same**.

Average Kinetic Energy is *Proportional* to *Absolute Temperature*

There are **two equations** for the **product** pV of a gas — the ideal gas equation (page 47), and the equation involving the mean square speed of the particles (page 48). You can **equate these** to get an expression for the **average kinetic energy**.

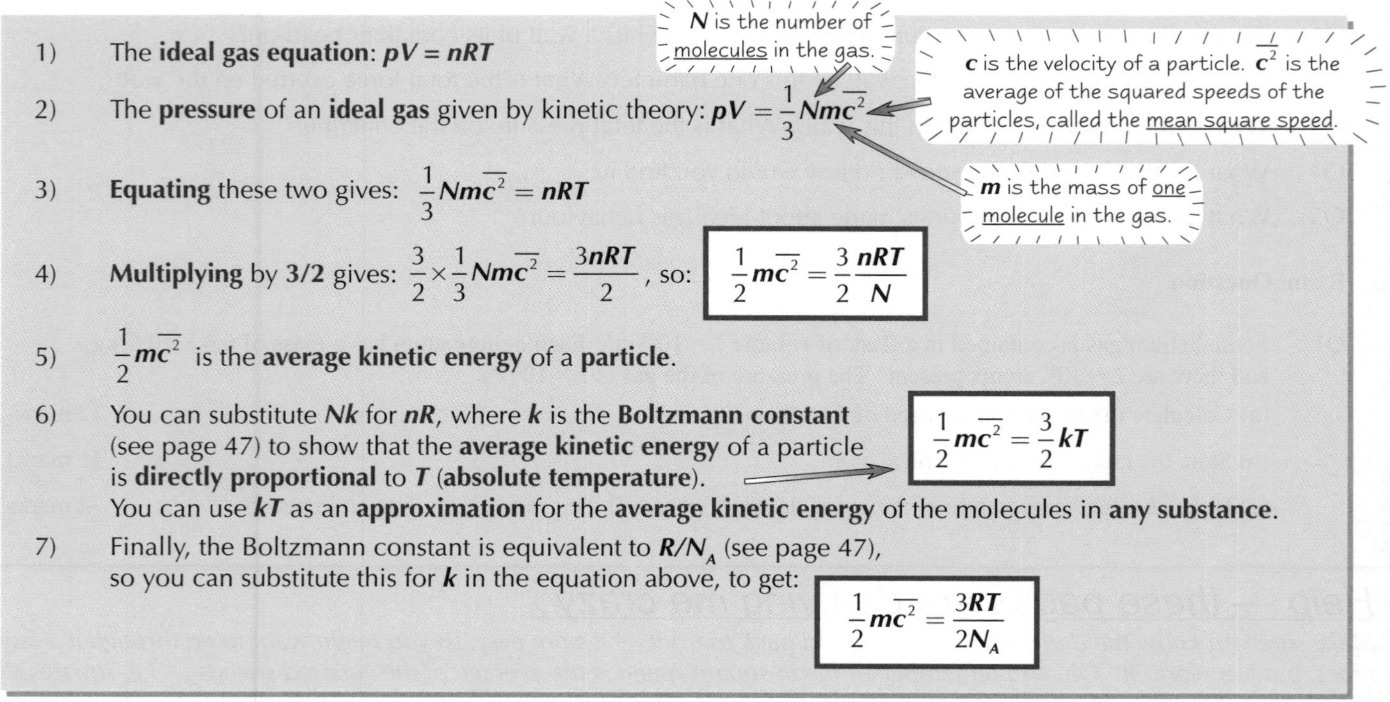

1) The **ideal gas equation:** $pV = nRT$
2) The **pressure** of an **ideal gas** given by kinetic theory: $pV = \frac{1}{3}Nm\overline{c^2}$
3) **Equating** these two gives: $\frac{1}{3}Nm\overline{c^2} = nRT$
4) **Multiplying** by **3/2** gives: $\frac{3}{2} \times \frac{1}{3}Nm\overline{c^2} = \frac{3nRT}{2}$, so: $\frac{1}{2}m\overline{c^2} = \frac{3}{2}\frac{nRT}{N}$
5) $\frac{1}{2}m\overline{c^2}$ is the **average kinetic energy** of a **particle**.
6) You can substitute Nk for nR, where k is the **Boltzmann constant** (see page 47) to show that the **average kinetic energy** of a particle is **directly proportional** to T (**absolute temperature**). $\frac{1}{2}m\overline{c^2} = \frac{3}{2}kT$
You can use kT as an **approximation** for the **average kinetic energy** of the molecules in **any substance**.
7) Finally, the Boltzmann constant is equivalent to R/N_A (see page 47), so you can substitute this for k in the equation above, to get: $\frac{1}{2}m\overline{c^2} = \frac{3RT}{2N_A}$

Energy and Temperature

Specific Heat Capacity is how much *Energy* it Takes to *Heat* Something

When you heat something, its particles get more **kinetic energy** and its **temperature** rises.

The **specific heat capacity** (*c*) of a substance is the amount of **energy** needed to **raise** the **temperature** of **1 kg** of the substance by **1 K** (or 1°C).

or put another way: **energy change = mass × specific heat capacity × change in temperature**

in symbols: $\Delta Q = mc\Delta\theta$ ← ΔT is sometimes used instead of $\Delta\theta$ for the change in temperature.

ΔQ is the energy change in J, m is the mass in kg and $\Delta\theta$ is the temperature change in K or °C. Units of **c** are J kg^{-1} K^{-1} or J kg^{-1} °C^{-1}.

Specific Latent Heat is the *Energy Needed* to *Change State*

To **melt** a **solid** or **boil or evaporate a liquid**, you need **energy** to **break the bonds** that hold the particles in place. The **energy** needed for this is called **latent heat**. The **larger** the **mass** of the substance, the **more energy** it takes to **change** its **state**. That's why the **specific latent heat** is defined per kg:

The **specific latent heat** (*l*) of **fusion** or **vaporisation** is the quantity of **thermal energy** required to **change the state** of **1 kg** of a substance.

which gives: **energy change = specific latent heat × mass of substance changed**

or in symbols: $\Delta Q = ml$ ← You'll usually see the latent heat of vaporisation written l_v and the latent heat of fusion written l_f.

Where ΔQ is the energy change in J and m is the mass in kg. The units of l are J kg^{-1}.

Practice Questions

Q1 Describe the changes in the distribution of gas particle speeds as the temperature of a gas increases.

Q2 What happens to the average kinetic energy of a particle if the temperature of a gas doubles?

Q3 Define specific heat capacity and specific latent heat.

Q4 Show that the thermal energy needed to heat 2 kg of water from 20 °C to 50 °C is ~250 kJ (c_{water} = 4180 J$kg^{-1}K^{-1}$).

Exam Questions

Q1 The mass of one mole of nitrogen molecules is 2.8×10^{-2} kg. There are 6.02×10^{23} molecules in one mole.

(a) Calculate the typical speed of a nitrogen molecule at 300 K. [4 marks]

(b) Explain why all the nitrogen molecules will not be moving at this speed. [2 marks]

Q2 Some air freshener is sprayed at one end of a room. The room is 8.0 m long and the temperature is 20 °C.

(a) Assuming the average freshener molecule moves at 400 ms^{-1}, how long will it take for a particle to travel directly to the other end of the room? [1 mark]

(b) The perfume from the air freshener only slowly diffuses from one end of the room to the other. Explain why this takes much longer than suggested by your answer to part (a). [2 marks]

Q3 A 2 kg metal cylinder is heated uniformly from 4.5 °C to 12.7 °C in 3 minutes.
The electric heater used is rated at 12 V, 7.5 A.
Assuming that heat losses were negligible, calculate the specific heat capacity of the metal. [3 marks]

Q4 A 3 kW electric kettle contains 0.5 kg of water already at its boiling point.
Neglecting heat losses, determine how long it will take to boil dry. (l_v (water) = 2.26×10^6 J kg^{-1}) [3 marks]

My specific eat capacity — 24 pies...

*Phew... there's a lot to take in here. Go back over it, step by step, and make sure you're comfortable using those equations. Interesting(ish) fact — it's the **huge** difference in SHC between the land and the sea that causes the monsoon in Asia.*

Optical Telescopes

Some optical telescopes use lenses (no, really), so first, here's a bit of lens theory...

Converging Lenses Bring Light Rays Together

1) **Lenses** change the **direction** of light rays by **refraction**.
2) Rays **parallel** to the **principal axis** of the lens converge onto a point called the **principal focus**. Parallel rays that **aren't** parallel to the principal axis converge somewhere else on the **focal plane** (see diagram).
3) The **focal length**, ***f***, is the distance between the **lens axis** and the **principal focus**.

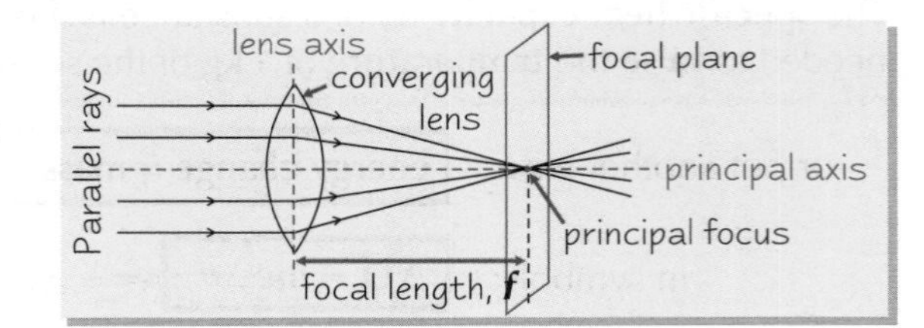

Images can be Real or Virtual

1) A **real image** is formed when light rays from an object are made to **pass through** another point in space. The light rays are **actually there**, and the image can be **captured** on a **screen**.
2) A **virtual image** is formed when light rays from an object **appear** to have come from another point in space. The light rays **aren't really where the image appears to be**, so the image **can't** be captured on a screen.
3) Converging lenses can form both **real** and **virtual** images, depending on where the object is. If the object is **further** than the **focal length** away from the lens, the image is **real**. If the object's **closer**, the image is **virtual**.
4) To work out where an image will appear, you can draw a **ray diagram**. You only need to draw **two rays** on a ray diagram: one **parallel** to the principal axis that passes through the **principal focus**, and one passing through the **centre** of the lens that **doesn't get bent**.

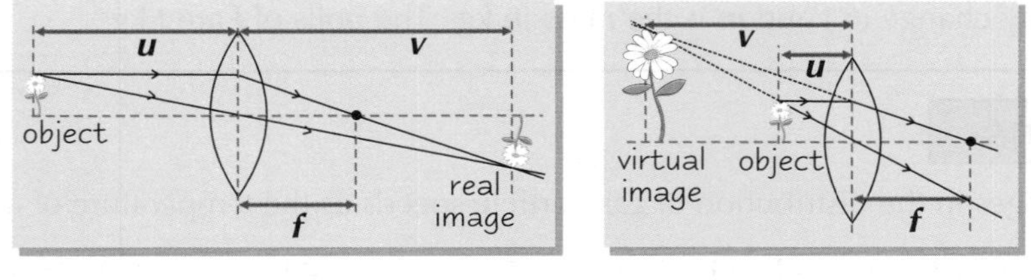

In the diagram, u = distance between object and lens axis, v = distance between image and lens axis (**positive** if image is **real**, **negative** if image is **virtual**), f = focal length.

5) The values u, v and f are related by the **lens equation**:

$$\frac{1}{f} = \frac{1}{u} + \frac{1}{v}$$

A Refracting Telescope uses Two Converging Lenses

1) The **objective lens** converges the rays from the object to form a **real image**.
2) The **eye lens** acts as a **magnifying glass** on this real image to form a **magnified virtual image**.

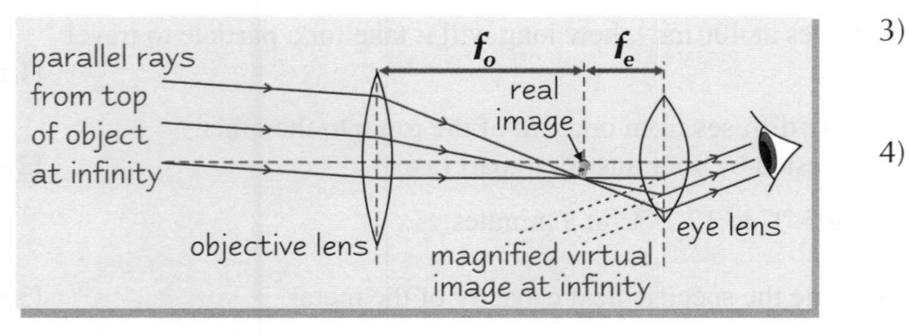

3) If you assume the object is at infinity, then the rays from it are **parallel**, and the real image is formed on the **focal plane**.
4) A **telescope** (in normal adjustment) is set up so that the **principal focus** of the **objective** lens is in the **same position** as the principal focus of the **eye** lens, so the **final magnified image** appears to be at **infinity**.
5) The **magnification**, ***M***, of the telescope can be calculated in terms of angles, or the focal length. The **angular magnification** is the **angle** subtended by the **image** θ_i over the **angle** subtended by the **object** θ_o at the eye:

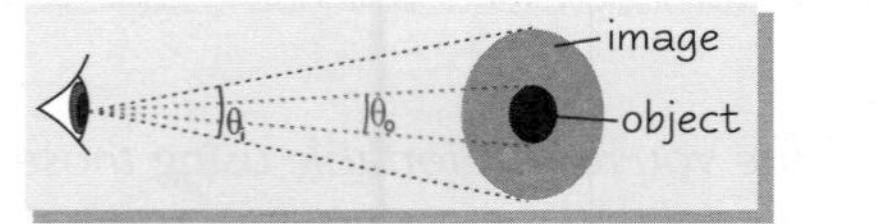

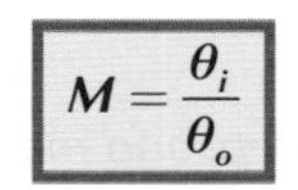

$$M = \frac{\theta_i}{\theta_o}$$

or in terms of **focal length** (with the telescope in normal adjustment as shown above):

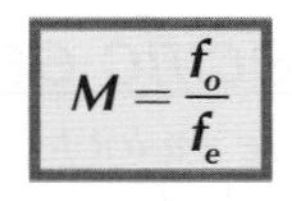

$$M = \frac{f_o}{f_e}$$

Optical Telescopes

A Reflecting Telescope uses a Concave Mirror and a Converging Lens

1) A **parabolic concave mirror** (the **primary mirror**) converges parallel rays from an object, forming a **real image**.
2) An **eye lens magnifies** the image as before.
3) The focal point of the mirror (where the image is formed) is **in front** of the mirror, so an arrangement needs to be devised where the observer doesn't **block out** the light. A set-up called the **Cassegrain arrangement**, which uses a **convex secondary mirror**, is a common solution to this problem.

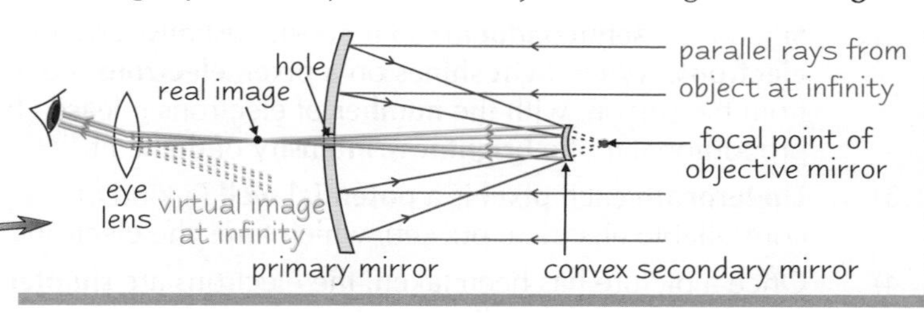

The Resolving Power of a Telescope — how much Detail you can See

1) The **resolving power** of an instrument is the **smallest angle** at which it can **distinguish** two points.

Two stars that can only just be distinguished

θ = resolving power

About half of the stars that we see in the night sky are actually collections of two or more stars. Our eyes see them as a single star since the angle between them is too small to resolve.

2) Resolution is limited by diffraction. If a beam of light passes through a circular **aperture**, then a **diffraction pattern** is formed. The central circle is called the **Airy disc** (see p.96 for an example of the pattern).
3) **Two** light sources can **just** be distinguished if the **centre** of the **Airy disc** from one source is **at least as far away** as the **first minimum** of the other source. This led to the **Rayleigh criterion**:

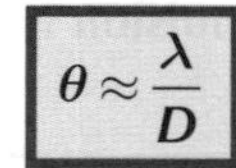

$$\theta \approx \frac{\lambda}{D}$$

where θ is the **minimum angle** that can be resolved in **radians**, λ is the **wavelength** of the light in **metres** and D is the **diameter** of the **aperture** in **metres**.

4) For **telescopes**, D is the diameter of the **objective lens** or the **objective mirror**. So **very large** lenses or mirrors are needed to see **fine detail**.

There are Big Problems with Refracting Telescopes

1) Glass refracts **different colours** of light by **different amounts** and so the image for each colour is in a slightly **different position**. This **blurs** the image and is called **chromatic aberration**.

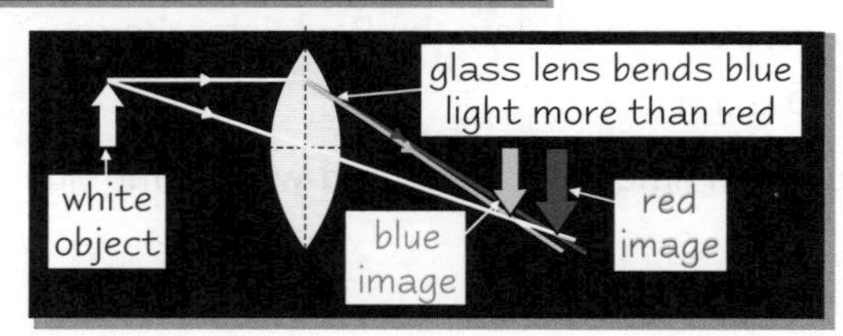

2) Any **bubbles** and **impurities** in the glass **absorb** some of the light, which means that **very faint** objects **aren't seen**. Building large lenses that are of a **sufficiently good quality** is **difficult** and **expensive**.
3) **Large lenses** are very **heavy** and can only be **supported** from their **edges**, so their **shape** can become **distorted**.
4) For a **large magnification**, the **objective lens** needs to have a **very long focal length**. This means that refracting telescopes have to be **very long**, leading to very **large** and **expensive buildings** needed to house them.

Reflecting Telescopes are Better than Refractors but they have Problems too

1) **Large mirrors** of **good quality** are much **cheaper** to build than large lenses. They can also be **supported** from **underneath** so they don't **distort** as much as lenses.
2) Mirrors don't suffer from **chromatic aberration** (see above) but can have **spherical aberration**:

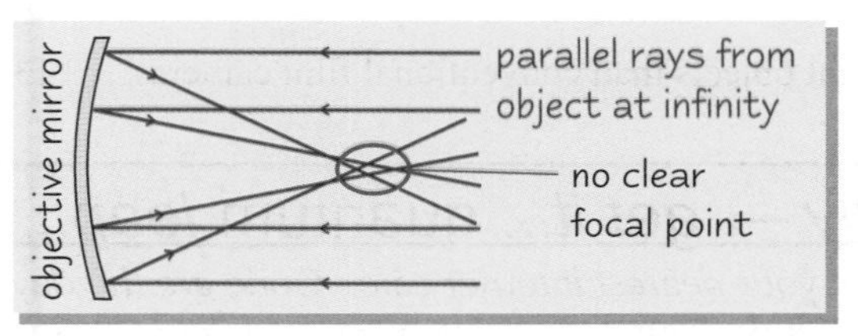

If the **shape** of the mirror isn't quite **parabolic**, parallel rays reflecting off different parts of the mirror do not all **converge** onto the same point.

When the **Hubble Space Telescope** was first launched it suffered from **spherical aberration**. They had to find a way round the problem before it could be used.

Optical Telescopes

Charge-Coupled Devices (CCDs) are Very Sensitive Light Detectors

1) CCDs are **silicon chips** about the size of a postage stamp, divided up into a grid of millions of **identical pixels**.
2) Silicon is a **semiconductor** so it doesn't usually have many **free electrons**. When **light** shines on a pixel, **electrons** are released from the silicon, with the **number** of electrons released being proportional to the **brightness/intensity** of the light.
3) **Underneath** each **pixel** is a **potential well** (a kind of controllable electrical bucket), which traps the electrons.
4) Once a picture has been taken, the electrons are **shunted** from **one potential well** to **another** so that they all come out **in sequence** from **one corner** of the CCD. (This is called an 'electrical bucket brigade', to use the lingo.)
5) This sequence can be converted into a **digital signal** and sent to computers **anywhere in the world**.

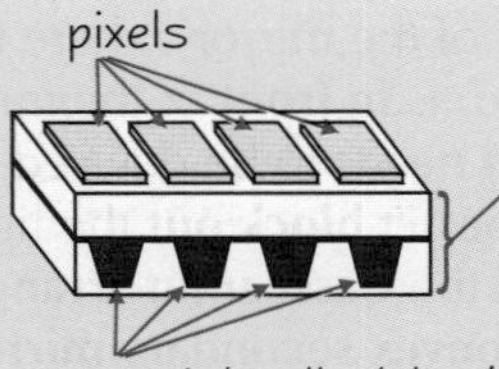

CCDs use Quantum Effects

1) **Quantum physics** tells us that EM radiation is formed in **discrete packets** of energy called **photons**.
2) The incoming photons release electrons in the silicon due to the **photoelectric effect**. (See p. 94.)
3) Electrons are released by **more than 70%** of the photons that hit a pixel, so the **quantum efficiency** of a CCD is greater than 70%. On average, a cell in the **eye** needs about **100 photons** before it responds and so has a quantum efficiency of about **1%**. The quantum efficiency of a **photographic emulsion** is about **4%**.

Practice Questions

Q1 Define the focal length of a converging lens.

Q2 Draw ray diagrams to show how an image is formed in a refracting and a reflecting (Cassegrain) telescope.

Q3 Explain resolving power and state the Rayleigh criterion.

Q4 What does CCD stand for?

Q5 What does quantum efficiency mean and what is the quantum efficiency of a CCD?

Exam Questions

Q1 (a) Define the *principal focus* and the *focal length* of a converging lens. [2 marks]

(b) An object was placed 0.20 m in front of a converging lens with a focal length of 0.15 m. Calculate how far behind the lens the image was formed. [3 marks]

(c) The object was placed 0.10 m in front of the same lens. Determine where the image was formed. [2 marks]

Q2 An objective lens with a focal length of 5.0 m and an eye lens with a focal length of 0.10 m are used in a refracting telescope.

(a) How far apart should the lenses be placed for the telescope to be in normal adjustment? [1 mark]

(b) Define angular magnification and calculate the angular magnification of this telescope. [2 marks]

Q3 (a) Describe the basic function of a CCD. [3 marks]

(b) Explain why CCDs are much better at taking pictures of faint objects than conventional film cameras. [3 marks]

CCDs were a quantum leap for astronomy — get it... quantum leap... *sigh*

With CCDs, you can get all the images you want from the comfort of your nearest internet café. Gone are the days of standing on a hill with a telescope and a flask hoping the sky clears before your nose turns black and falls off. Shame.

Non-Optical Telescopes

Some telescopes don't use visible light — they use radio waves, IR, UV or X-rays instead — read on to learn more...

Radio Telescopes are **Similar** to **Optical** Telescopes in Some Ways

1) The most obvious feature of a radio telescope is its **parabolic dish**.
This works in exactly the same way as the **objective mirror** of an **optical reflecting** telescope.
2) An **antenna** is used as a detector at the **focal point** instead of an eye or camera in an optical telescope, but there is **no equivalent** to the **eye lens**.

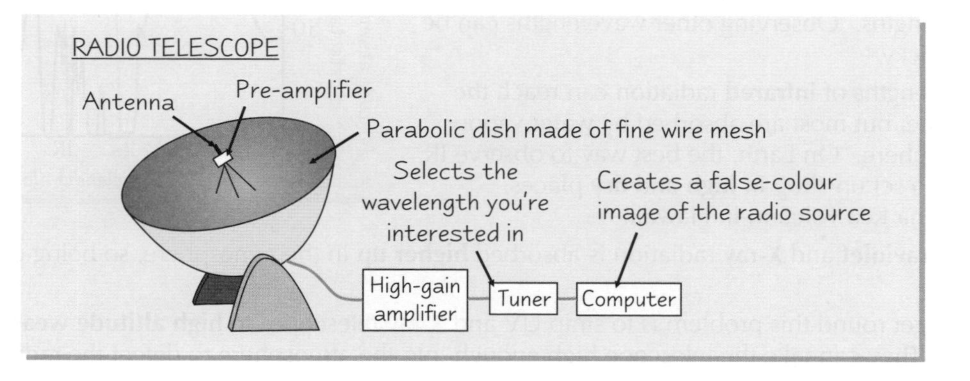

3) Most radio telescopes are **manoeuvrable**, allowing the source of the waves to be **tracked** (in the same way as optical telescopes). The telescope moves with the source, stopping it 'slipping out of view' as the Earth rotates.

Radio Waves have a Much **Longer Wavelength** than Light...

1) The **wavelengths** of **radio waves** are about a **million times longer** than the wavelengths of **light**.
2) The **Rayleigh criterion** (see p. 53) gives the **resolving power** of a telescope as $\theta \approx \lambda/D$.
3) So for a radio telescope to have the **same resolving power** as an optical telescope, its dish would need to be a **million times bigger** (about the size of the UK for a decent one). The **resolving power** of a radio telescope is **worse** than the **unaided eye**.

Radio astronomers get around this by **linking** lots of telescopes together.
Using some nifty computer programming, their data can be combined to form a **single image**. This is equivalent to one **huge dish** the size of the **separation** of the telescopes.
Resolutions **thousands** of times better than optical telescopes can be achieved this way.

...so **Radio** Telescopes aren't as **Fiddly** to Make as Optical Reflectors

1) Instead of a **polished mirror**, a **wire mesh** can be used since the long wavelength radio waves don't notice the gaps. This makes their **construction** much **easier** and **cheaper** than optical reflectors.
2) The **shape** of the dish has to have a **precision** of about **λ/20** to avoid **spherical aberration** (see page 53).
So the dish does not have to be **anywhere near as perfect** as a mirror.
3) Unlike an optical telescope, a radio telescope has to **scan across** the radio source to **build up** the **image**.

Non-Optical Telescopes

The **Atmosphere Blocks** *Certain* **EM Wavelengths**

1) One of the big problems with doing astronomy on Earth is trying to look through the atmosphere.
2) Our atmosphere only lets **certain wavelengths** of **electromagnetic radiation** through and is **opaque** to all the others. The graph shows how the **transparency** of the atmosphere varies with **wavelength**.
3) We can use **optical** and **radio** telescopes on the surface of the Earth because the atmosphere is **transparent** to these wavelengths. Observing other wavelengths can be a bit more tricky.
4) A few wavelengths of **infrared** radiation can reach the Earth's surface, but most are absorbed by water vapour in the atmosphere. On Earth, the best way to observe IR radiation is to set up shop in **high** and **dry** places, like the Mauna Kea volcano in Hawaii.

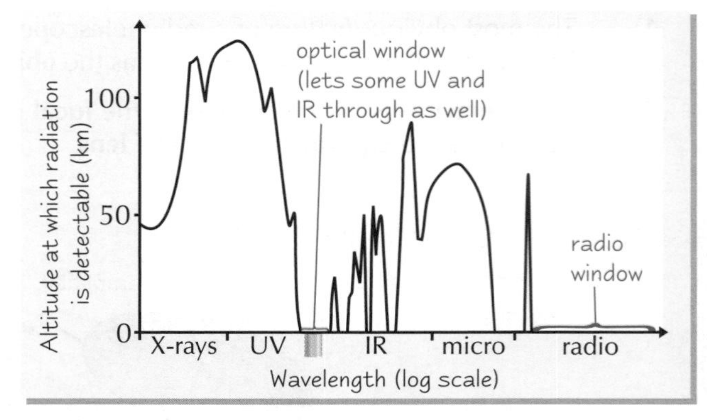

5) But most **ultraviolet** and **X-ray** radiation is absorbed **higher up** in the atmosphere, so being on a mountain doesn't help.
6) One way to get round this problem is to strap UV and X-ray telescopes to **high altitude weather balloons** or **aeroplanes**. They can take the telescope high enough into the atmosphere to detect the radiation.
7) The ideal situation is to get your telescope **above the atmosphere** altogether, by launching it into **space** and setting it in orbit around the Earth.

IR *and* **UV** *Telescopes have a* **Very Similar Structure** *to* **Optical** *Telescopes*

1) Infrared and ultraviolet telescopes are very similar to optical reflecting telescopes. They use the same **parabolic mirror** set-up to focus the radiation onto a detector.
2) In both cases, **CCDs** (see p 54) or **special photographic paper** are used as the radiation detectors, just as in optical telescopes.
3) The **longer** the **wavelength** of the radiation, the **less** it's affected by imperfections in the mirror (see previous page). So the mirrors in **infrared** telescopes **don't** need to be as perfectly shaped as in optical telescopes. But the mirrors in **UV** telescopes have to be even **more** precisely made.

IR telescopes have the added problem that they produce their **own** infrared radiation due to their **temperature**. They need to be **cooled** to very low temperatures using liquid helium, or refrigeration units.

X-ray Telescopes *have a* **Different Structure** *from Other Telescopes*

1) X-rays don't reflect off surfaces in the same way as most other EM radiation. Usually X-ray radiation is either **absorbed** by a material or it **passes straight through** it.

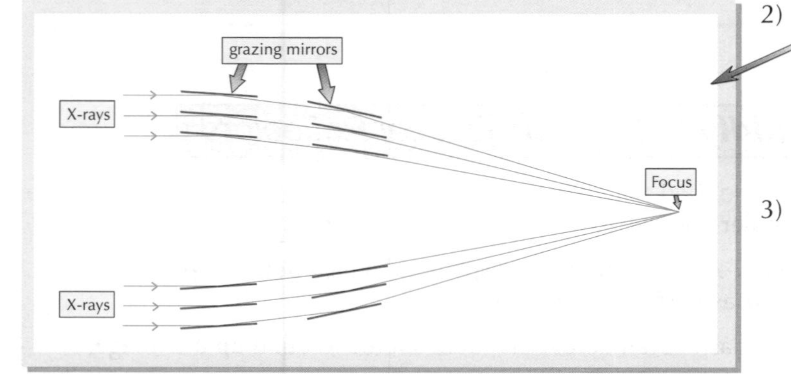

2) X-rays **do** reflect if they just **graze** a mirror's surface though. By having a series of **nested mirrors**, you can gradually alter the direction of X-rays enough to bring them to a **focus** on a detector. This type of telescope is called a **grazing telescope**.
3) The X-rays can be detected using a modified **Geiger counter** or a **fine wire mesh**. Modern X-ray telescopes such as the XMM-Newton telescope use highly sensitive X-ray **CCD** cameras.

Non-Optical Telescopes

Different Telescopes have Different **Resolving** and **Collecting Powers**

The **RESOLVING POWER** of a telescope is limited by two main factors:

1) The **Rayleigh criterion** (see page 53):
This depends on the **wavelength** of the radiation and the **diameter** of the objective mirror or dish.
So, for the **same size** of dish, a UV telescope has a much better resolving power than a radio telescope.

2) The quality of the **detector**:
Just like in digital cameras, the resolving power of a telescope is limited by the resolving power of the detector. That can be how many **pixels** there are on a CCD, or for a wire mesh X-ray detector, how **fine** the wire mesh is.

The **COLLECTING POWER** of a telescope is proportional to its **collecting area**.

1) A **bigger dish** or **mirror** collects **more energy** from an object in a given time.
This gives a **more intense image**, so the telescope can observe **fainter** objects.

The bigger the dish, the greater the collecting power. Mmm....

2) The **collecting power** (energy collected per second) is proportional to the area:

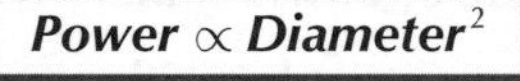

3) For a **radio**, **optical**, **UV** or **IR** telescope, this is the area of the objective mirror or dish.

4) For **X-ray** telescopes, it's the size of the **opening** through which X-rays can enter the telescope.
In general, X-ray telescopes have a much **smaller collecting power** than other types of telescope.

Practice Questions

Q1 Why do radio telescopes tend to have poor resolving powers?

Q2 Why is it easier to make a parabolic dish for a radio telescope than it is to make a mirror for an optical telescope?

Q3 Why don't astronomers install UV and X-ray telescopes on the top of mountains?

Exam Questions

Q1 Describe and explain the differences in resolving and collecting powers between radio and UV telescopes. [4 marks]

Q2 In 1983, the IRAS satellite observed the entire sky in infrared wavelengths. The satellite was kept at a temperature of 2 K by a reservoir of liquid helium which cooled the satellite by evaporation.

(a) Why did the satellite need to be kept at such a low temperature? [2 marks]

(b) Some infrared telescopes are on the surface of the Earth. What is special about their location? [1 mark]

Q3 (a) Many X-ray and UV telescopes are housed on satellites that orbit high above the Earth's atmosphere. Where else are X-ray and UV telescopes positioned? Explain why this is necessary. [2 marks]

(b) Describe and explain the major differences between the mirrors in X-ray and UV telescopes. [3 marks]

Q4 (a) How is the collecting power of a telescope related to its objective diameter? [1 mark]

(b) The Arecibo radio telescope has a dish diameter of 300 m. The Lovell radio telescope has a dish diameter of 76 m. Calculate the ratio of their collecting powers. [2 marks]

Power is proportional to diameter^2? Bring on the cakes...

If you can't observe the radiation you want to from Earth, just strap your telescope to a rocket and blast it into space. Sounds easy enough till you remember it's going to be reeeally hard to repair if anything goes wrong.

Distances and Magnitude

There are a couple of ways to classify stars — the first is by luminosity, using the magnitude scale.

The **Luminosity** of a Star is the **Total Energy** Emitted **per Second**

1) Stars can be **classified** according to their luminosity — that is, the **total** amount of energy emitted in the form of electromagnetic radiation **each second** (see p.60).
2) The **Sun's** luminosity is about 4×10^{26} W (luminosity is measured in watts, since it's a sort of power). The **most luminous** stars have a luminosity about a **million** times that of the Sun.
3) The **intensity**, I, of an object that we observe is the power **received** from it per unit area **at Earth**. This is the effective **brightness** of an object.

Apparent Magnitude, m, is based on how **Bright** things **Appear** from **Earth**

1) The **brightness** of a star in the night sky depends on **two** things — its **luminosity** and its **distance from us** (if you ignore weather and light pollution, etc.). So the **brightest** stars will be **close** to us and have a **high luminosity**.
2) The ancient Greeks invented a system where the very **brightest** stars were given an **apparent magnitude** of **1** and the **dimmest** stars an apparent magnitude of **6**, with other levels catering for the stars in between.
3) In the 19th century, the scale was redefined using a strict **logarithmic** scale:

   A **magnitude 1** star has an **intensity 100 times** greater than a **magnitude 6** star.

   This means a difference of **one magnitude** corresponds to a difference in **intensity** of $\mathbf{100^{1/5}}$ **times**. So a magnitude 1 star is about **2.5 times brighter** than a magnitude 2 star.
4) At the same time, the range was **extended** in **both directions** with the very brightest objects in the sky having **negative apparent magnitude**.

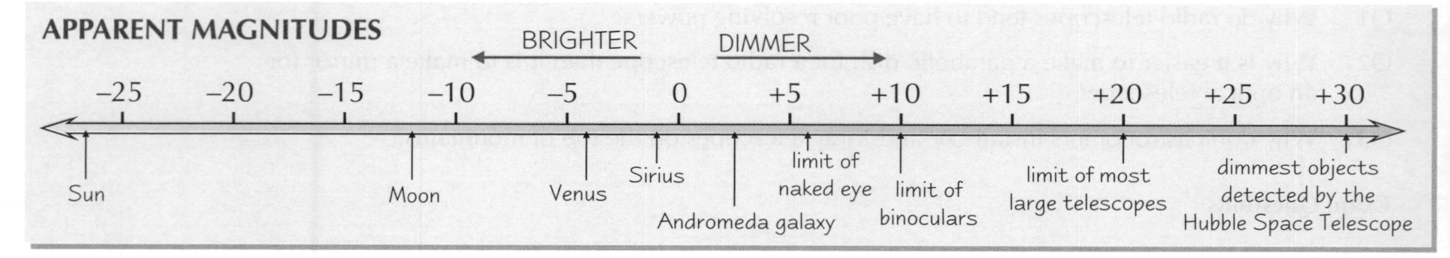

5) The **apparent magnitude**, m, is related to the **intensity**, I, by the following formula:

$$m = -2.5 \log I + \text{constant}$$

Note that log is $\log_{10}$, not the natural logarithm ln. And that 2.5 is exactly 2.5 — not $100^{1/5}$.

The **Distance** to **Nearby Stars** can be Measured in **Parsecs**

1) Imagine you're in a **moving car**. You see that (stationary) objects in the **foreground** seem to be **moving faster** than objects in the **distance**. This **apparent change in position** is called **parallax**.
2) Parallax is measured in terms of the **angle of parallax**. The **greater** the **angle**, the **nearer** the object is to you.
3) The distance to **nearby stars** can be calculated by observing how they **move relative** to **very distant stars** when the Earth is in **different parts** of its **orbit**. This gives a **unit** of distance called a **parsec** (**pc**).

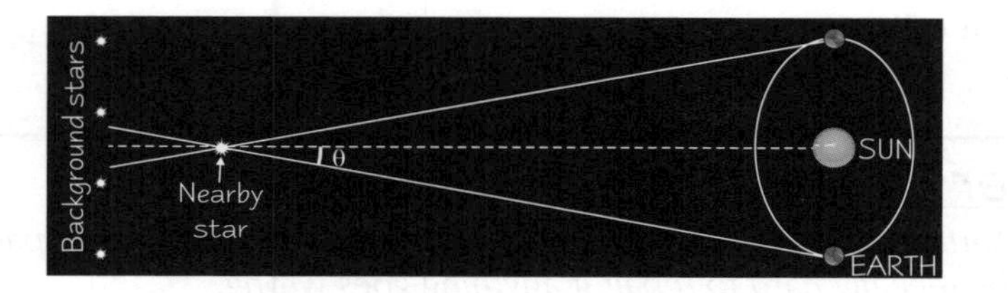

A star is exactly **one parsec (pc)** away from Earth if the **angle of parallax**,

$$\theta = \mathbf{1\ arcsecond} = \left(\frac{1}{3600}\right)^{\circ}$$

Distances and Magnitude

*Absolute Magnitude, M, is based only on the **Luminosity** of the Star*

1) The **absolute magnitude** of a star or galaxy, ***M***, does not depend on its distance from Earth. It is defined as what its apparent magnitude **would be** if it were **10 parsecs** away from Earth.
2) The relationship between ***M*** and ***m*** is given by the following formula:

$$m - M = 5\log\left(\frac{d}{10}\right)$$

where ***d*** is the distance in parsecs

If you know the absolute magnitude of a star, you can use this equation to calculate its **distance** from Earth. This is really handy, since the distance to most stars is **too big** to measure using parallax (see previous page).

This method uses objects like **supernovae** that are known as **standard candles**. Standard candles are objects that you can calculate the luminosity of **directly**. So, if you find a supernova within a galaxy, you can work out how far that galaxy is from us. This is how the **Hubble constant** was worked out (see p. 68).

Distances** in the Solar System are Often Measured in **Astronomical Units (AU)

The **parsec** is only one measurement used in **astrophysics** — luckily the others you need to know are much **simpler**.

1) From **Copernicus** onwards, astronomers were able to work out the **distance** the **planets** are from the Sun **relative** to the Earth, using **astronomical units** (AU). But they could not work out the **actual distances**.

   One **astronomical unit** (AU) is defined as the **mean distance** between the **Earth** and the **Sun**.

2) The **size** of the AU (1.50×10^{11} m) wasn't known until 1769 — when it was carefully **measured** during a **transit of Venus** (when Venus passed between the Earth and the Sun).

*Another Measure of Distance is the **Light-Year (ly)***

1) All **electromagnetic waves** travel at the **speed of light**, **c**, in a vacuum ($\mathbf{c} = 3.00 \times 10^8$ ms^{-1}).

   The **distance** that electromagnetic waves travel through a vacuum in **one year** is called a **light-year** (**ly**).

2) If we see the light from a star that is, say, **10 light-years away** then we are actually seeing it as it was **10 years ago**. The further away the object is, the further **back in time** we are actually seeing it.
3) **1 ly** is equivalent to about **63 000 AU**.

Practice Questions

Q1 What is the relationship between apparent magnitude and intensity?

Q2 What is the equation that links apparent magnitude, absolute magnitude and distance?

Q3 Give three units of distance used in astrophysics. Explain the meaning of each one.

Exam Questions

Q1 Define the *absolute magnitude* of a star. [2 marks]

Q2 Calculate the absolute magnitude of the Sun given that the Sun's apparent magnitude is –27. [1 pc = 2 × 10^5 AU] [4 marks]

Q3 The star Sirius has an apparent magnitude of –1.46 and an absolute magnitude of +1.4. The star Canopus has an apparent magnitude of –0.72 and an absolute magnitude of –5.5.

(a) State which of the two stars appears brighter from Earth. [1 mark]

(b) Calculate the distance of Canopus from Earth. [3 marks]

*Logs — the cheap and easy alternative pet...**

The magnitude scale is a pretty weird system, but like with a lot of astronomy, the old ways have stuck. Remember — the lower the number, the brighter the object. The definition of absolute magnitude is a bit random as well — I mean, why ten parsecs? Ours not to reason why, ours but to... erm... learn it. (Doesn't have quite the same ring does it.)

* Too young to remember Twin Peaks...? Ah well, never mind.

Stars as Black Bodies

Now they're telling us the Sun's black. Who writes this stuff?

A Black Body is a Perfect Absorber and Emitter

1) Objects emit **electromagnetic radiation** due to their **temperature**. At everyday temperatures this radiation lies mostly in the **infrared** part of the spectrum (which we can't see) — but heat something up enough and it will start to **glow**.
2) **Pure black** surfaces emit radiation **strongly** and in a **well-defined way**. We call it **black body radiation**.
3) A black body is defined as:

   A body that **absorbs all wavelengths** of electromagnetic radiation (that's why it's called a **black** body) and can **emit all wavelengths** of electromagnetic radiation.

4) To a reasonably good approximation **stars** behave as **black bodies** and their black body radiation produces their **continuous spectrum**.
5) The graph of **intensity** against **wavelength** for black body radiation varies with **temperature**, as shown in the graph:

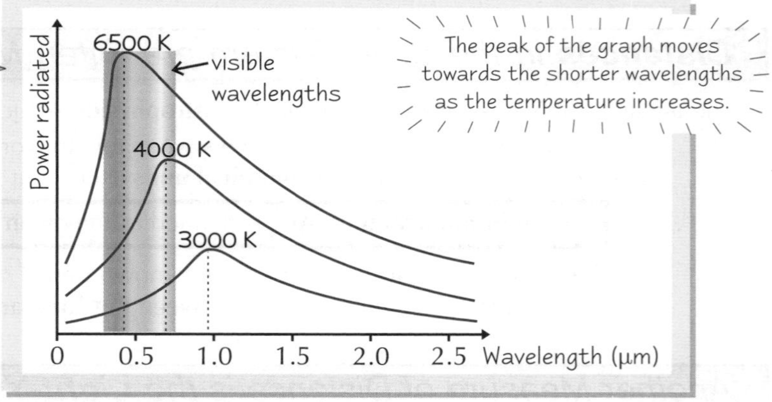

The Peak Wavelength gives the Temperature

1) For each temperature, there is a **peak** in the black body curve at a wavelength called the **peak wavelength**, λ_{max}.
2) λ_{max} is related to the **temperature** by **Wien's displacement law**:

$$\lambda_{max} T = 0.0029 \text{ m·K}$$

where T is the temperature in kelvin and m·K is a metre-kelvin.

The Luminosity of a Star Depends on its Temperature and Surface Area

1) The **luminosity** of a star is the **total energy** it emits **per second** and is related to the **temperature** of the star and its **surface area**.
2) The luminosity is proportional to the **fourth power** of the star's **temperature** and is **directly proportional** to the **surface area**. This is **Stefan's law**:

$$L = \sigma A T^4$$

where L is the luminosity of the star (in W), A is its surface area (in m^2), T is its surface temperature (in K) and σ (a little Greek "sigma") is Stefan's constant.

3) Measurements give Stefan's constant as $\sigma = 5.67 \times 10^{-8}$ $Wm^{-2}K^{-4}$.
4) From **Earth**, we can measure the **intensity** of the star. The intensity is the **power** of radiation **per square metre**, so as the radiation spreads out and becomes **diluted**, the intensity **decreases**. If the energy has been emitted from a **point** or a **sphere** (like a star, for example) then it obeys the **inverse square law**:

$$I = \frac{L}{4\pi d^2}$$

where L is the luminosity of the star (in W), and d is the distance from the star.

Stars as Black Bodies

You Can Put the **Equations Together** to **Solve Problems**

Example The star Sirius B has a surface area of 4.1×10^{13} m^2 and produces a black body spectrum with a peak wavelength of 115 nm. The intensity of the light from Sirius B when it reaches Earth is 1.12×10^{-11} Wm^{-2}. How long does the light from Sirius B take to reach Earth? *($\sigma = 5.67 \times 10^{-8}$ Wm^{-2}K^{-4}, $c = 3.0 \times 10^8$ ms^{-1})*

First, find the temperature of Sirius B:

$$\lambda_{max} T = 0.0029 \text{ m·K, so } T = 0.0029 \div \lambda_{max} = 0.0029 \div 115 \times 10^{-9} = 25\ 217 \text{ K.}$$

Now, you can use Stefan's law to find the luminosity:

$$L = \sigma A T^4 = (5.67 \times 10^{-8}) \times (4.1 \times 10^{13}) \times 25\ 217^4 = 9.4 \times 10^{23} \text{ W}$$

Then use $I = \frac{L}{4\pi d^2}$ to find the distance of Sirius B from Earth:

$$d = \sqrt{\frac{L}{4\pi I}} = \sqrt{\frac{9.4 \times 10^{23}}{4 \times \pi \times 1.12 \times 10^{-11}}} = \sqrt{6.68 \times 10^{23}} = 8.2 \times 10^{16} \text{m}$$

Finally, use $c = d \div t$ to find the time taken $t = d \div c = 8.2 \times 10^{16} \div 3 \times 10^8 = 273\ 333\ 333$ s $\approx$ **8.7 years**

It's Hard to get **Accurate** Measurements

1) **Wien's displacement law**, **Stefan's law** and the **inverse square law** can all be used to work out various **properties** of stars. This needs very **careful measurements**, but our **atmosphere** mucks up the results.
2) It only lets through **certain wavelengths** of **electromagnetic radiation** — **visible** light, most **radio** waves, **very near infrared** and a bit of **UV**. It's **opaque** to the rest.
3) And then there are things like **dust** and **man-made light pollution** to contend with. Observatories are placed at **high altitudes**, well away from **cities**, and in **low-humidity** climates to minimise the problem. The best solution, though, is to send up **satellites** that can take measurements **above** the atmosphere.
4) Our **detectors** don't do us any favours either. The **measuring devices** that astronomers use aren't perfect since their **sensitivity** depends on the **wavelength**.
5) For example, **glass absorbs UV** light but is **transparent** to **visible light**, so any instruments that use glass affect UV readings straight off.
6) All you can do about this is choose the best materials for what you want to measure, and then **calibrate** your instruments really carefully.

Practice Questions

Q1 What is Wien's displacement law and what is it used for?

Q2 What is the relationship between luminosity, surface area and temperature?

Q3 Why are accurate measurements of black body radiation difficult on the Earth's surface?

Exam Questions

Q1 A star has a surface temperature of 4000 K and the same luminosity as the Sun (3.9×10^{26} W).

(a) Which radiation curve represents this star — X, Y or Z? Explain your answer. [2 marks]

(b) Calculate the star's surface area. [2 marks]

Q2 The star Procyon A, which has a luminosity of 2.3×10^{27} W, produces a black body spectrum with a peak wavelength of 436 nm.
Calculate the surface area of Procyon A. [4 marks]

Astronomy — theories, a bit of guesswork and a whole load of calibration...

Astronomy isn't the most exact of sciences, I'm afraid. The Hubble Space Telescope's improved things a lot, but try and get a look at some actual observational data. Then look at the error bars — they'll generally be about the size of a house.

Spectral Classes and the H-R Diagram

As well as classifying stars by luminosity (the magnitude scale, p.58), they can be classified by colour.

The **Visible** Part of **Hydrogen's Spectrum** is called the **Balmer Series**

1) The lines in **emission** and **absorption spectra** occur because electrons in an atom can only exist at certain well-defined **energy levels**.
2) In **atomic hydrogen**, the electron is usually in the **ground state** (**n** = 1), but there are lots of energy levels (**n** = 2 to **n** = ∞ — called excitation levels) that the electron **could** exist in if it was given more energy.

> The wavelengths corresponding to the **visible bit** of hydrogen's spectrum are caused by electrons moving from **higher energy levels** to the **first excitation level** (**n = 2**).
> This leads to a series of **lines** called the **Balmer series.**

The **Strengths** of the **Spectral Lines** Show the **Temperature** of a Star

1) For a **hydrogen absorption line** to occur in the **visible** part of a star's spectrum, electrons in the hydrogen atoms already need to be in the **n = 2** state.
2) This happens at **high temperatures**, where **collisions** between the atoms give the electrons extra energy.
3) If the temperature is **too high**, though, the majority of the electrons will reach the **n = 3** level (or above) instead, which means there won't be so many Balmer transitions.
4) So the **intensity** of the Balmer lines depends on the **temperature** of the star.
5) For a particular intensity of the Balmer lines, **two temperatures** are possible. Astronomers get around this by looking at the **absorption lines** of **other atoms** and **molecules** as well.

The **Relative Strength** of Absorption Lines gives the **Spectral Class**

Well... quite.

1) For historical reasons the stars are classified into:

   spectral classes: O (hottest), **B, A, F, G, K and M**

   Use a **mnemonic** to remember the order. The standard one is the rather non-PC **'Oh Be A Fine Girl, Kiss Me'**.
2) The graph shows how the **intensity** of the visible spectral lines changes with **temperature**:

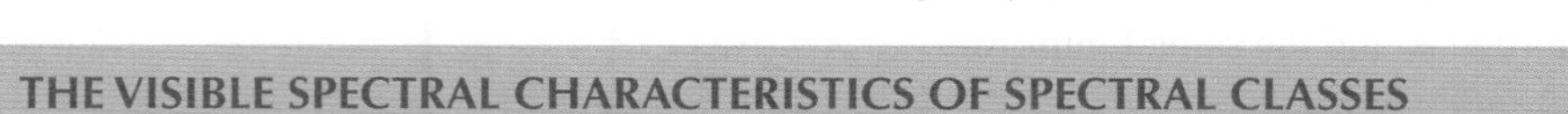

A quick note on the temperature axis: These diagrams, and the H-R diagram (next page), tend to be drawn with spectral class along the horizontal axis. The relationship between spectral class and temperature isn't linear or logarithmic.

THE VISIBLE SPECTRAL CHARACTERISTICS OF SPECTRAL CLASSES

- **O** — **Blue** stars: temperature **25 000 – 50 000 K.** The strongest spectral lines are **helium ion** and **helium atom** absorptions, since these need a really high temperature. They have weak **hydrogen Balmer** lines too.
- **B** — **Blue** stars: **11 000 – 25 000 K.** These spectra show strong **helium atom** and **hydrogen** absorptions.
- **A** — **Blue-white** stars: **7500 – 11 000 K.** Visible spectra are governed by very strong Balmer **hydrogen** lines, but there are also some **metal ion** absorptions.
- **F** — **White** stars: **6000 – 7500 K.** These spectra have strong **metal ion** absorptions.
- **G** — **Yellow-white** stars: **5000 – 6000 K.** These have both **metal ion** and **metal atom** absorptions.
- **K** — **Orange** stars: **3500 – 5000 K.** At this temperature, spectral lines are mostly from neutral **metal atoms.**
- **M** — **Red** stars: **< 3500 K. Molecular band** absorptions from compounds like **titanium oxide** are present in the spectra of these stars, since they're cool enough for molecules to form.

Spectral Classes and the H-R Diagram

Absolute Magnitude vs Temperature/Spectral Class — the H-R diagram

1) Independently, Hertzsprung and Russell noticed that a plot of **absolute magnitude** (see p. 59) against **temperature** (or **spectral class**) didn't just throw up a random collection of stars but showed **distinct areas**.

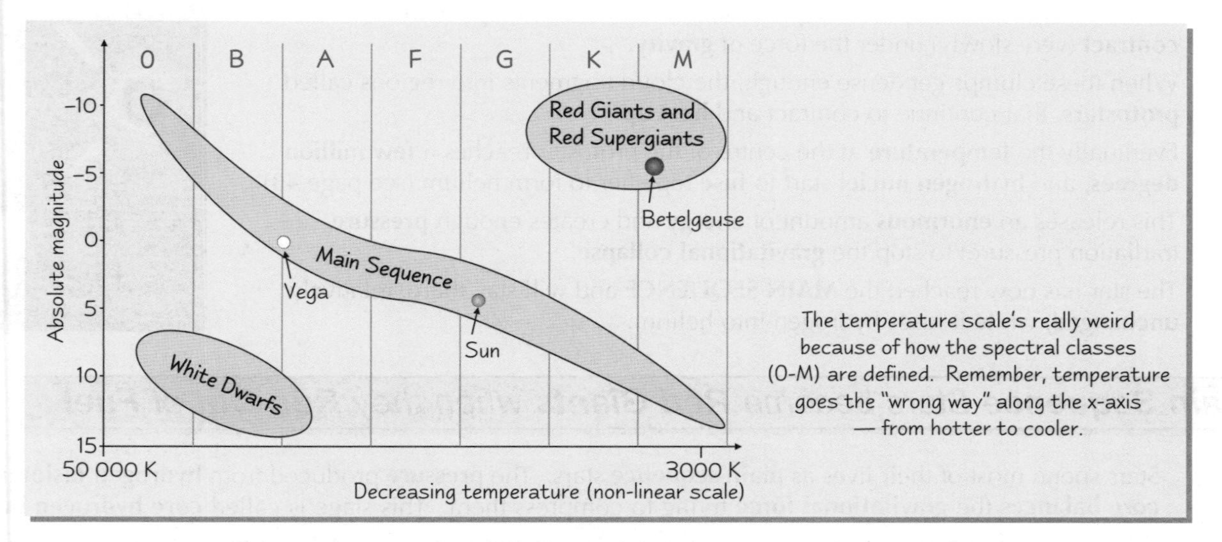

2) The **long, diagonal band** is called the **main sequence**. Main sequence stars are in their long-lived **stable phase** where they are fusing **hydrogen** into **helium**. The Sun is a main sequence star.

3) Stars that have a **high luminosity** and a relatively **low surface temperature** must have a **huge** surface area because of Stefan's law (page 60). These stars are called **red giants** and are found in the **top-right** corner of the H-R diagram. Red giants are stars that have **moved off** the main sequence and fusion reactions other than hydrogen to helium are also happening in them.

4) Stars that have a **low luminosity** but a **high temperature** must be very **small**, again because of Stefan's law. These stars are called **white dwarfs** and are about the size of the Earth. They lie in the **bottom-left** corner of the H-R diagram. White dwarfs are stars at the **end** of their lives, where all of their fusion reactions have stopped and they are just **slowly cooling down**.

Practice Questions

Q1 Why does hydrogen have to be at a particular temperature before Balmer absorption lines are seen?

Q2 List the spectral classes in order of decreasing temperature and outline their spectral characteristics.

Q3 What is an H-R diagram and what are the three main groups of stars that emerge when the diagram is plotted?

Exam Questions

Q1 The spectral classes of stars can be identified by examining the lines in their absorption spectra.

(a) Explain how temperature affects the strength of the Balmer lines in stellar absorption spectra. [3 marks]

(b) State the two spectral classes of star in which strong Balmer lines are observed. [2 marks]

(c) Describe the visible spectral characteristics and temperature of a star in class F. [3 marks]

Q2 The spectra of K and M stars have absorption bands corresponding to energy levels of molecules. Explain why this only occurs in the lowest temperature stars. [2 marks]

Q3 Draw the basic features of an H-R diagram, indicating where you would find main sequence stars, red giants and white dwarfs. [5 marks]

'Ospital Bound — A Furious Girl Kicked Me...

Spectral classes are another example of astronomers sticking with tradition. The classes used to be ordered alphabetically by the strength of the Balmer lines. When astronomers realised this didn't quite work, they just fiddled around with the old classes rather than coming up with a sensible new system. Just to make life difficult for people like you and me.

Stellar Evolution

Stars go through several different stages in their lives and move around the H-R diagram as they go *(see p. 63).*

Stars *Begin as Clouds of* ***Dust*** *and* ***Gas***

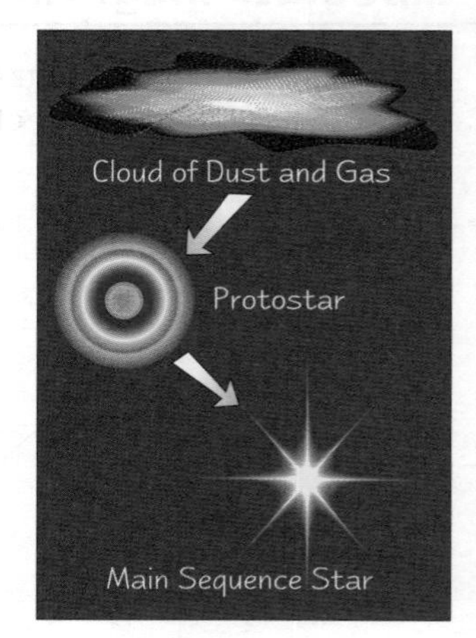

1) Stars are born in a **cloud** of **dust** and **gas**, most of which was left when previous stars blew themselves apart in **supernovae**. The denser clumps of the cloud **contract** (very slowly) under the force of **gravity**.
2) When these clumps get dense enough, the cloud fragments into regions called **protostars**, that continue to contract and **heat up**.
3) Eventually the **temperature** at the centre of the protostar reaches a **few million degrees**, and **hydrogen nuclei** start to **fuse** together to form helium (see page 43).
4) This releases an **enormous** amount of **energy** and creates enough **pressure** (radiation pressure) to stop the **gravitational collapse**.
5) The star has now reached the **MAIN SEQUENCE** and will stay there, relatively **unchanged**, while it fuses hydrogen into helium.

Main Sequence *Stars become* ***Red Giants*** *when they* ***Run Out*** *of* ***Fuel***

1) Stars spend most of their lives as **main sequence** stars. The **pressure** produced from **hydrogen fusion** in their **core balances** the **gravitational force** trying to compress them. This stage is called **core hydrogen burning**.
2) When the **hydrogen** in the **core** runs out nuclear fusion **stops**, and with it the **outward pressure stops**. The core **contracts** and **heats up** under the **weight** of the star. The outer layers **expand** and **cool**, and the star becomes a **RED GIANT**.
3) The material **surrounding** the core still has **plenty of hydrogen**. The **heat** from the contracting **core** raises the **temperature** of this material enough for the hydrogen to **fuse**. This is called **shell hydrogen burning**. (Very low-mass stars stop at this point. They use up their fuel and slowly fade away...)
4) The core continues to contract until, eventually, it gets **hot** enough and **dense** enough for **helium** to **fuse** into **carbon** and **oxygen**. This is called **core helium burning**. This releases a **huge** amount of energy, which **pushes** the **outer layers** of the star further outwards.
5) When the **helium** runs out, the carbon-oxygen core **contracts again** and heats a **shell** around it so that helium can fuse in this region — **shell helium burning**.

Low Mass *Stars (like the Sun)* ***Eject*** *their Shells, leaving behind a* ***White Dwarf***

1) In low-mass stars, the **carbon-oxygen core isn't hot enough** for any further **fusion** and so it continues to **contract** under its own **weight**. Once the core has shrunk to about **Earth-size**, **electrons** exert enough pressure (**electron degeneracy pressure**) to stop it collapsing any more (fret not — you don't have to know how).
2) The **helium shell** becomes more and more **unstable** as the core contracts. The star **pulsates** and **ejects** its outer layers into space as a **planetary nebula**, leaving behind the dense core.
3) The star is now a very **hot**, **dense solid** called a **WHITE DWARF**, which will simply **cool down** and **fade away**.

High Mass *Stars have a* ***Shorter Life*** *and a more* ***Exciting Death***

1) Even though stars with a **large mass** have a **lot of fuel**, they use it up **more quickly** and don't spend so long as main sequence stars.
2) When they are **red giants** the **'core burning to shell burning'** process can continue beyond the fusion of helium, building up layers in an **onion-like structure**. For **really massive** stars this can go all the way up to **iron**.
3) Nuclear fusion **beyond iron** isn't **energetically favourable**, though, so once an iron core is formed then very quickly it's goodbye star.
4) The star explodes cataclysmically in a **SUPERNOVA**, leaving behind a **NEUTRON STAR** or (if the star was massive enough) a **BLACK HOLE**.

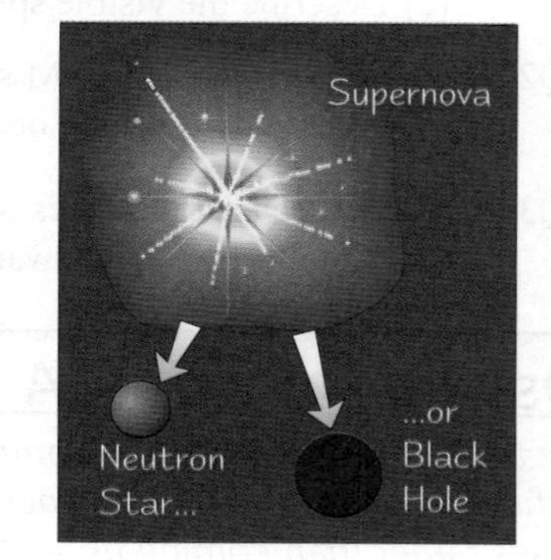

Stellar Evolution

Massive Stars go out with a Bit of a Bang

1) When the core of a star runs out of fuel, it starts to **contract** — forming a white dwarf core.
2) If the star is **massive enough**, though, **electron degeneracy** can't stop the core contracting. This happens when the mass of the core is more than **1.4 times** the mass of the **Sun**.
3) The electrons get **squashed** onto the atomic **nuclei**, combining with protons to form **neutrons** and **neutrinos**.
4) The core suddenly collapses to become a **NEUTRON STAR**, which the outer layers then **fall** onto.
5) When the outer layers **hit** the surface of the **neutron star** they **rebound**, setting up huge **shockwaves**, ripping the star apart and causing a **supernova**. The light from a supernova can briefly outshine an **entire galaxy**.

Neutron stars are incredibly **dense** (about 4×10^{17} kgm^{-3}).
They're **very small**, typically about 20 km across, and they can **rotate very fast** (up to 600 times a second).
They emit **radio waves** in two beams as they rotate. These beams sometimes sweep past the Earth and can be observed as **radio pulses** rather like the flashes of a lighthouse. These rotating neutron stars are called **PULSARS**.

Neutron Stars are Weird but Black Holes are a Lot Weirder

1) If the **core** of the star is more than **3 times** the **Sun's mass**, the **neutrons** can't withstand the gravitational forces.
2) There are **no known mechanisms** left to stop the core collapsing to an **infinitely dense** point called a **singularity**. At that point, the **laws of physics** break down completely.
3) Up to a certain distance away (called the **Schwarzschild radius**) the gravitational pull is **so strong** that nothing, not even **light**, can escape its grasp. The **boundary** of this region is called the **event horizon**.

The Schwarzschild radius is the distance at which the escape velocity is the speed of light
An object moving at the **escape velocity** has **just enough kinetic energy** to overcome the black hole's gravitational field.

From Newton's law of gravitation we get $\frac{1}{2}mv^2 = \frac{GMm}{r}$ where m = mass of object, M = mass of black hole, v = velocity of object, r = distance from centre of black hole, $G = 6.67 \times 10^{-11}$ Nm^{2}kg^{-2}

Dividing through by m and making r the subject gives: $r = \frac{2GM}{v^2}$

By replacing v with the speed of light, c, you get the Schwarzschild radius, R_S: $R_S = \frac{2GM}{c^2}$.

This derivation is a bit of a fudge (although it gives the right answer, as it happens) — Newton's law of gravity doesn't quite work in intense gravitational fields, so Einstein's general relativity should be used instead (the maths is way too hard to go into here though).

Practice Questions

Q1 Outline how the Sun was formed, how it will evolve and how it will die.

Q2 Describe a white dwarf and a neutron star. What are the main differences between them?

Exam Questions

Q1 Outline the main differences between the evolution of high mass and low mass stars, starting from when they first become main sequence stars. [6 marks]

Q2 (a) What is meant by the Schwarzschild radius of a black hole? [2 marks]

(b) Calculate the Schwarzschild radius for a black hole that has a mass of 6×10^{30} kg. [2 marks]

Live fast — die young...

The more massive a star, the more spectacular its life cycle. The most massive stars burn up the hydrogen in their core so quickly that they only live for a fraction of the Sun's lifetime — but when they go, they do it in style.

The Doppler Effect and Redshift

Everyone's heard of the Big Bang theory — well here's some evidence for it.

The Doppler Effect — the Motion of a Wave's Source Affects its Wavelength

1) You'll have experienced the Doppler effect **loads of times** with **sound waves**.
2) Imagine an ambulance driving past you. As it moves **towards you** its siren sounds **higher-pitched**, but as it **moves away**, its **pitch** is **lower**. This change in **frequency** and **wavelength** is called the **Doppler shift**.
3) The frequency and the wavelength **change** because the waves **bunch together** in **front** of the source and **stretch out behind** it. The **amount** of stretching or bunching together depends on the **velocity** of the **source**.
4) When a **light source** moves **away** from us, the wavelengths of its light become **longer** and the frequencies become lower. This shifts the light towards the **red** end of the spectrum and is called **redshift**.
5) When a light source moves **towards** us, the **opposite** happens and the light undergoes **blueshift**.
6) The amount of redshift or blueshift, z, is determined by the following formula:

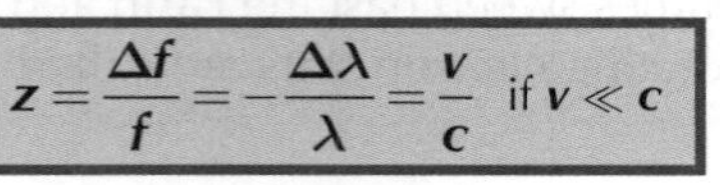

$$z = \frac{\Delta f}{f} = -\frac{\Delta \lambda}{\lambda} = \frac{v}{c} \quad \text{if } v \ll c$$

λ is the emitted wavelength, f is the emitted frequency, $\Delta\lambda$ and Δf are the differences between the observed and emitted wavelengths/frequencies, v is the velocity of the source in the observer's direction and c is the speed of light. ($v \ll c$ means "v is much less than c".)

7) The way cosmologists tend to look at this stuff, the galaxies aren't actually moving **through space** away from us. Instead, **space itself** is expanding and the light waves are being **stretched** along with it. This is called **cosmological redshift** to distinguish it from **redshift** produced by sources that **are** moving through space.
8) The same formula works for both types of redshift as long as v is much less than c. If v is close to the speed of light, you need to use a nasty, relativistic formula instead (you don't need to know that one).

The Red Shift of Galaxies is Strong Evidence for the HBB

1) The **spectra** from **galaxies** (apart from a few very close ones) all show **redshift** — the **characteristic spectral lines** of the elements are all at a **longer wavelength** than you would expect. This shows they're all **moving apart**.
2) Hubble realised that the **speed** that **galaxies moved away** from us depended on **how far** they were away. This led to the idea that the Universe started out **very hot** and **very dense** and is currently **expanding**.

THE HOT BIG BANG THEORY (HBB): the Universe started off **very hot** and **very dense** (perhaps as an **infinitely hot, infinitely dense** singularity) and has been **expanding** ever since.

Redshift is Used to Study Spectroscopic Binary Stars

1) About half of the stars we observe are actually **two stars** that orbit each other. Many of them are too far away from us to be **resolved** with **telescopes** but the **lines** in their **spectra** show a binary star system. These are called **spectroscopic binary stars**.
2) By observing how the **absorption lines** in the spectrum change with **time** the **orbital period** can be calculated:

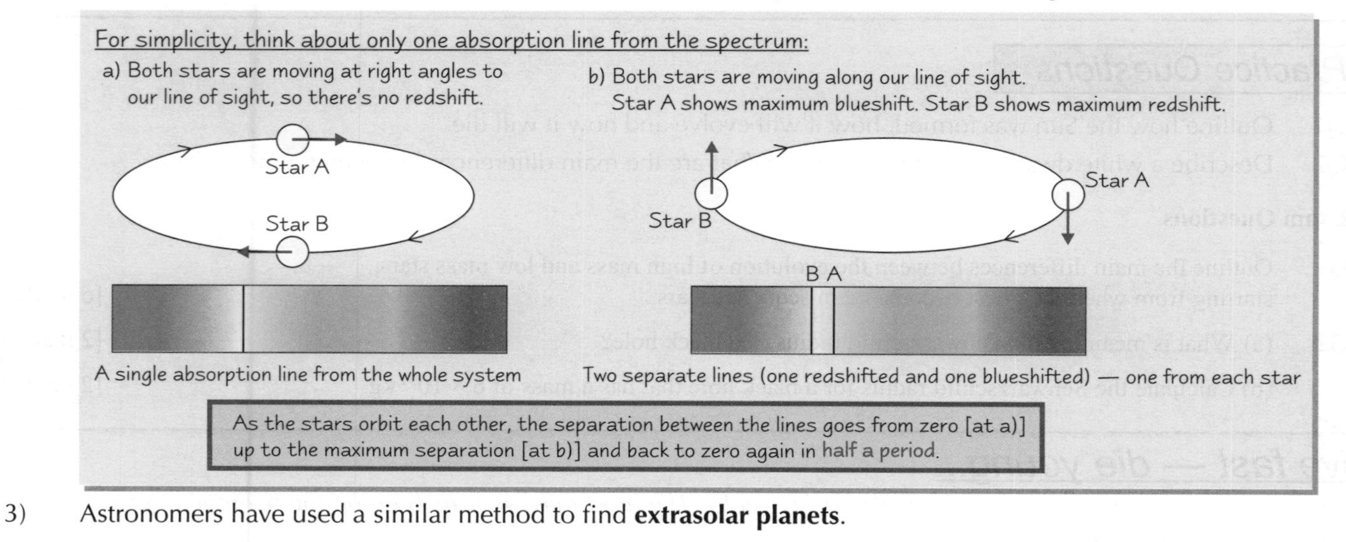

3) Astronomers have used a similar method to find **extrasolar planets**.

The Doppler Effect and Redshift

The Doppler effect explains quasars too. Read on.

Quasars — Quasi-Stellar Objects

1) **Quasars** were discovered in the late 1950s and were first thought to be **stars in our galaxy**.
2) The puzzling thing was that their spectra were **nothing like** normal stars. They sometimes shot out **jets** of material, and many of them were very active **radio sources**.
3) The 'stars' produced a **continuous spectrum** that was nothing like a black body radiation curve and instead of absorption lines, there were **emission lines** of elements that astronomers **had not seen before**.
4) However, these lines looked strangely familiar and in 1963 Maarten Schmidt realised that they were simply the **Balmer series** of hydrogen (see p.62) but **redshifted** enormously.

Quasars are a Very Long Way Away so they must be Very Bright

1) This **huge redshift** suggests they're a **huge distance away** (see next page) — in fact, the **most distant** objects seen.
2) The measured redshifts give us distances of **billions of light years**.
Using the **inverse square law** for intensity (see p. 60) gives an idea of just how **bright** quasars are:

Example A quasar has the same intensity as a star 20 000 ly away with the same luminosity as the Sun (4×10^{26} W). Its redshift gives a distance of 1×10^{10} ly. Calculate its luminosity.

$I_{quasar} = I_{star}$ so they cancel out of the equation.

$$L \propto Id^2 \Rightarrow \frac{L_{quasar}}{L_{star}} = \frac{d_{quasar}^2}{d_{star}^2} \Rightarrow L_{quasar} = L_{star} \cdot \frac{d_{quasar}^2}{d_{star}^2} = 4\times10^{26} \cdot \frac{1\times10^{20}}{4\times10^{8}} = 1\times10^{38}\ \text{W}$$

That's bright — about **10 times** the **luminosity** of the **entire Milky Way galaxy**!

3) And there's very good evidence to suggest that quasars are only about the size of the **Solar System**.
4) Let me run that past you again. **That's the power of a trillion Suns from something the size of the Solar System**.
5) These numbers caused a lot of controversy in the astrophysics community — they seemed crazy. Many astrophysicists thought there must be a more reasonable explanation. But then evidence for the distance of quasars came when **sensitive CCD** equipment detected the fuzzy cloud of **a galaxy around a quasar**.

The current consensus is that a quasar is a very powerful **galactic nucleus** — a huge **black hole** about 100 million times the mass of the Sun at the centre of a distant galaxy. (**All** galaxies are thought to have these 'supermassive' black holes at their centres).
This black hole is surrounded by a doughnut shaped mass of **whirling gas** falling into it, which produces the light. In the same way as a pulsar (see p.65), magnetic fields produce jets of radiation streaming out from the poles. The black hole must consume the mass of about **10 Suns per year** to produce the energy observed.

Practice Questions

Q1 What is the Doppler effect? Write down the formula for the redshift and blueshift of light.

Q2 Explain how the spectra of binary stars can be used to calculate their period of orbit.

Exam Questions

Q1 The spectra of three objects have been taken. What can you deduce from each of the following?

(a) The absorption lines from object A have been shifted towards the blue end of the spectrum. [1 mark]

(b) The absorption lines from object B oscillate either side of their normal position in the spectrum with a period of two weeks. [2 marks]

(c) The wavelength of the hydrogen alpha line in object C's spectrum is 667.83 nm. In the laboratory, the wavelength of the same line is measured as 656.28 nm. [3 marks]

Q2 (a) State one piece of evidence that suggests quasars are a very long distance away. [1 mark]

(b) Use the concept of the inverse square law to suggest why quasars must be very bright. [2 marks]

(c) Describe the main features of a quasar according to the current theory. [2 marks]

Long ago, in a galaxy far, far away — there was a radio-loud, supermassive black hole with a highly luminous arc...

Quasars are weird. There's still some disagreement in the astrophysics community about what they are. There's even some evidence (not generally accepted) that quasars are much nearer than the redshift suggests and are just moving very quickly.

The Big Bang Model of the Universe

Right, we're moving on to the BIG picture now — we all like a bit of cosmology...

The Universe is The Same in Every Direction

When you read that all the **galaxies** in the Universe are **moving away** from the **Earth** (see p.66 and below), it's easy to imagine that the Earth is at the **centre of the Universe**, or that there's something really **special** about it. **Earth** is special to us because we **live here** — but on a **universal scale**, it's just like any other lump of rock.

1) The **demotion** of **Earth** from anything special is taken to its logical conclusion with the **cosmological principle**...

> **COSMOLOGICAL PRINCIPLE**: on a **large scale** the Universe is **homogeneous** (every part is the same as every other part) and **isotropic** (everything looks the same in every direction) — so it doesn't have a **centre**.

2) Until the **1930s**, cosmologists believed that the Universe was **infinite** in both **space** and **time** (that is, it had always existed) and **static**. This seemed the **only way** that it could be **stable** using **Newton's law** of gravitation. Even **Einstein modified** his theory of **general relativity** to make it consistent with the **Steady-State Universe**.

Hubble Realised that the Universe is Expanding

1) The **spectra** from **galaxies** (apart from a few very close ones) all show **redshift**. The amount of **redshift** gives the **recessional velocity** — how fast the galaxy is moving away (see page 66).
2) A plot of **recessional velocity** against **distance** (found using standard candles — see p.59) showed that they were **proportional**, which suggests that the Universe is **expanding**. This gives rise to **Hubble's law**:

$$v = H_0 d$$

v = recessional velocity in **kms^{-1}**,
d = distance in **Mpc** and
H_0 = **Hubble's constant** in **$kms^{-1}Mpc^{-1}$**.

3) Since distance is very difficult to measure, astronomers disagree on the value of H_0. It's generally accepted that H_0 lies somewhere between 50 $kms^{-1}Mpc^{-1}$ and 100 $kms^{-1}Mpc^{-1}$.
4) The **SI unit** for H_0 is s^{-1}. To get H_0 in SI units, you need v in ms^{-1} and d in m (1 Mpc = 3.09×10^{22} m).

The Expanding Universe gives rise to the Hot Big Bang Model

1) The Universe is **expanding** and **cooling down** (because it's a closed system). So further back in time it must have been **smaller** and **hotter**. If you trace time back **far enough**, you get a **Hot Big Bang** (see page 66).
2) Since the Universe is **expanding uniformly** away from **us** it seems as though we're at the **centre** of the Universe, but this is an **illusion**. You would observe the **same thing** at **any point** in the Universe.

Hot big bang — the ultimate firework display?

The Age and Observable Size of the Universe Depend on H_0

1) If the Universe has been **expanding** at the **same rate** for its whole life, the **age** of the Universe is $t = 1/H_0$ (time = distance/speed). This is only an estimate though — see below.
2) Unfortunately, since no one knows the **exact value** of H_0 we can only guess the Universe's age. If $H_0 = 75$ **$kms^{-1}Mpc^{-1}$**, then the age of the Universe $\approx 1/(2.4 \times 10^{-18}\ s^{-1}) = 4.1 \times 10^{17}$ s = **13 billion years**.
3) The **absolute size** of the Universe is **unknown** but there is a limit on the size of the **observable Universe**. This is simply a **sphere** (with the Earth at its centre) with a **radius** equal to the **maximum distance** that **light** can travel during its **age**. So if $H_0 = 75$ **$kms^{-1}Mpc^{-1}$** then this sphere will have a radius of **13 billion light years**.

> **THE RATE OF EXPANSION HASN'T BEEN CONSTANT**
>
> All the **mass** in the Universe is attracted together by **gravity**. This attraction tends to **slow down** the rate of expansion of the Universe. It's thought that the expansion **was** decelerating until about 5 billion years ago.
>
> But in the late 90s, astronomers found evidence that the expansion is now **accelerating**. Cosmologists are trying to explain this acceleration using **dark energy** — a type of energy that fills the whole of space. There are various theories of what this dark energy **is**, but it's really hard to test them.

The Big Bang Model of the Universe

Cosmic Microwave Background Radiation — More Evidence for the HBB

1) The Hot Big Bang model predicts that loads of **electromagnetic radiation** was produced in the **very early Universe**. This radiation should **still** be observed today (it hasn't had anywhere else to go).
2) Because the Universe has **expanded**, the wavelengths of this cosmic background radiation have been **stretched** and are now in the **microwave** region.
3) This was picked up **accidentally** by Penzias and Wilson in the 1960s.

Properties of the Cosmic Microwave Background Radiation (CMBR)

1) In the late 1980s a satellite called the **Cosmic Background Explorer** (**COBE**) was sent up to have a **detailed look** at the radiation.
2) It found a **perfect blackbody spectrum** corresponding to a **temperature** of **2.73 K** (see page 60).
3) The radiation is largely **isotropic** and **homogeneous**, which confirms the cosmological principle (see page 68).
4) There are **very tiny fluctuations** in temperature, which were at the limit of COBE's detection. These are due to tiny energy-density variations in the early Universe, and are needed for the initial '**seeding**' of galaxy formation.
5) The background radiation also shows a **Doppler shift**, indicating the Earth's motion through space. It turns out that the **Milky Way** is rushing towards an unknown mass (the **Great Attractor**) at over a **million miles an hour**.

Another Bit of Evidence is the Amount of Helium in the Universe

1) The HBB model also explained the **large abundance of helium** in the Universe (which had puzzled astrophysicists for a while).
2) The early Universe had been very hot, so at some point it must have been hot enough for **hydrogen fusion** to happen. This means that, together with the theory of the synthesis of the **heavier elements** in stars, the **relative abundances** of all of the elements can be accounted for.

Practice Questions

Q1 What is Hubble's law? How can it be used to find the age of the Universe?

Q2 What is the cosmic background radiation?

Q3 How do the relative amounts of hydrogen and helium in the Universe provide evidence for the HBB model?

Exam Questions

Q1 (a) State Hubble's law, explaining the meanings of all the symbols. [2 marks]

(b) What does Hubble's law suggest about the nature of the Universe? [2 marks]

(c) Assume $H_0 = +50$ kms^{-1}Mpc^{-1} (1 Mpc = 3.09×10^{22} m).

i) Calculate H_0 in SI units. [2 marks]

ii) Calculate an estimate of the age of the Universe, and hence the size of the observable Universe. [3 marks]

Q2 (a) A certain object has a redshift of 0.37. Estimate the speed at which it is moving away from us. [2 marks]

(b) Use Hubble's law to estimate the distance (in light years) that the object is from us.
(Take $H_0 = 2.4 \times 10^{-18}$ s^{-1}, 1 ly = 9.5×10^{15} m.) [2 marks]

(c) With reference to the speed of the object, explain why your answers to a) and b) are estimates. [1 mark]

Q3 Describe the main features of the cosmic background radiation and explain why its discovery was considered strong evidence for the Hot Big Bang model of the Universe.
The quality of your written answer will be assessed in this question. [7 marks]

My Physics teacher was a Great Attractor — everyone fell for him...

The simple Big Bang model doesn't actually work — not quite, anyway. There are loads of little things that don't quite add up. Modern cosmologists are trying to improve the model using a period of very rapid expansion called inflation.

Physics of the Eye

*The eyes contain **converging lenses**, which focus light rays to form images. Page 52 tells you all about lenses — make sure you're confident with that before you start. Here are the two big equations you need to know:*

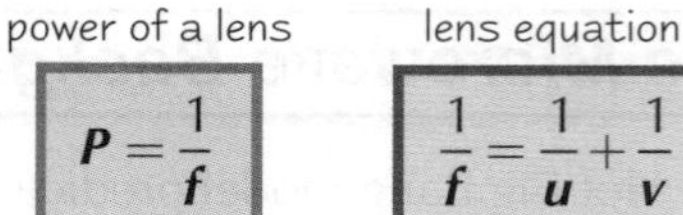

You Need to Know the Basic **Structure** of the **Eye**

1) The **cornea** is a **transparent** 'window' with a **convex** shape, and a **high refractive index**. The cornea does most of the eye's focusing.
2) The **aqueous humour** is a **watery** substance that lets light pass through the pupil to the lens.
3) The **iris** is the coloured part of the eye. It consists of **radial** and **circular muscles** that control the size of the **pupil** — the hole in the middle of the iris. This regulates the intensity of light entering the eye.
4) The **lens** acts as a **fine focus** and is controlled by the **ciliary muscles**. When the ciliary muscles **contract**, tension is released and the lens takes on a **fat**, more **spherical** shape. When they **relax**, the **suspensory ligaments** pull the lens into a **thin**, **flatter** shape.
5) The **vitreous humour** is a **jelly-like** substance that keeps the eye's shape.
6) Images are formed on the **retina**, which contains **light-sensitive cells** called **rods** and **cones** (see below).
7) The **yellow spot** is a particularly sensitive region of the retina. In the centre of the yellow spot is the **fovea**. This is the part of the retina with the highest concentration of **cones**.
8) The **optic nerve** carries signals from the rods and cones to the **brain**.

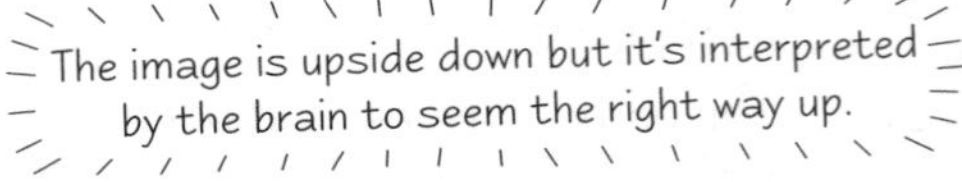

The **Eye** is an **Optical Refracting System**

1) The **far point** is the **furthest distance** that the eye can focus comfortably. For normally sighted people that's **infinity**. When your eyes are focusing at the far point, they're **'unaccommodated'**. The **near point** is the **closest distance** that the eye can focus on. For young people it's about 9 cm.
2) The **cornea** and **aqueous humour** act as a **fixed converging lens** with a **power** of about **41 D**.
3) The power of the eye's **lens** itself is about **18 D** when **unaccommodated**. By changing shape, it can increase to about **29 D** in young people to view objects at the **near point**.
4) You can **add together** the **powers** of the cornea, aqueous humour and lens. That means you can think of the eye as a **single converging lens** of **power 59 D** at the far point. This gives a **focal length** of **1.7 cm**.
5) When looking at nearer objects, the eye's power **increases**, as the lens changes shape and the **focal length decreases** — but the distance between the lens and the image, ***v***, stays the same, at 1.7 cm.

Power ≈ 67 D, f = 1.5 cm
u = 12.8 cm (from lens equation)

The **Retina** has **Rods and Cones**

1) **Rods** and **cones** are cells at the back of the **retina** that respond to **light**. Light travels **through the retina** to the rods and cones at the back.
2) Rods and cones all contain chemical **pigments** that **bleach** when **light** falls on them. This bleaching stimulates the cell to send signals to the **brain** via the **optic nerve**.
3) The cells are **reset** (i.e. unbleached) by enzymes using **vitamin A** from the blood.
4) There's only **one** type of **rod** but there are **three** types of **cone**, which are sensitive to **red**, **green** and **blue** light.

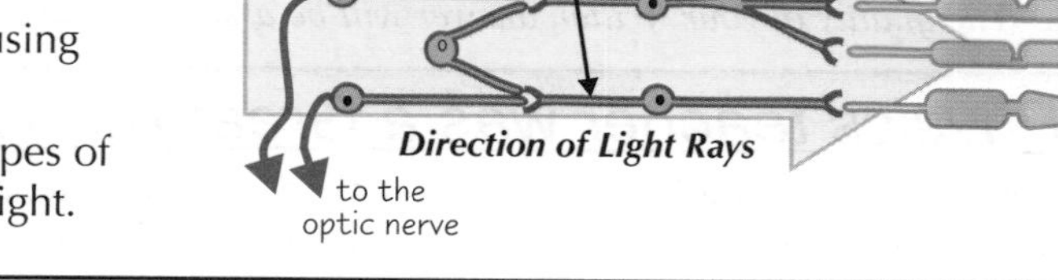

Physics of the Eye

The Cones let you See in Full Colour

1) The red, green and blue **cones** each absorb a **range of wavelengths**.
2) The eye is **less responsive** to blue light than to red or green, so blues often look dimmer.
3) The brain receives signals from the three types of cone and interprets their **weighted relative strengths** as **colour**...

Example
Yellow light produces almost equal responses from the red and green cones.
Yellow light can therefore be 'faked' by combining red and green light of almost equal intensity — the electrical signal from the retina will be the same and the brain interprets it as 'yellow'.

Relative absorption
total
400 500 600 700
wavelength (nm)

Weighted absorption
blue
green
red
300 350 400 450 500 550 600 650 700
Wavelength (nm)

4) **Any** colour can be produced by **combining** different intensities of **red**, **green** and **blue** light. Colour televisions work like this.

You Need Good Spatial Resolution to See Details

1) Two objects can only be distinguished from each other if there's **at least one rod** or **cone between** the light from each of them. Otherwise the brain can't **resolve** the two objects and it 'sees' them as one.

Too close, brain 'sees'

Just resolved, brain 'sees'

2) **Spatial resolution** is **best** at the **yellow spot** — the **cones** are very **densely packed** here and each cone always has its **own nerve fibre**. There are **no rods in the yellow spot**, though. This means that in **dim light**, when **cones don't work**, resolution is best slightly off the direct line of sight, where the **rods** are more **densely packed**.
3) Away from the yellow spot, resolution is much worse. The light-sensitive cells are **not** as **densely packed** and the rods **share nerve fibres** — there are up to 600 rods per fibre at the edges of the retina.

Persistence of Vision means you Don't See Rapid Flickering

1) **Nerve impulses** from the eye take about a fifth of a second to **decay**. So a very dim light flashing faster than **five times per second** (5 Hz) seems to be **on continuously**. This is called **flicker fusion**.
2) At **higher light intensity**, more nerve cells are 'firing' so a **higher frequency** is needed for flicker fusion to occur.
3) Cinema and TV rely on **persistence of vision** to give the illusion of **smooth** rather than **jerky movements**.

Practice Questions

Q1 Draw a ray diagram to show a young person's eye focusing at the near point. Assume that the eye's total power is 70 D. Mark on your diagram the distances $\boldsymbol{u}$, $\boldsymbol{v}$ and $\boldsymbol{f}$.

Q2 Describe the differences and similarities between rods and cones.

Q3 Sketch a graph showing how the cone cells in the retina respond to different wavelengths of light.

Q4 What is meant by 'persistence of vision'? Give a situation where it is useful.

Exam Questions

Q1 The power of an unaccommodated eye is 60 D.
(a) Determine the image distance, $\boldsymbol{v}$, when the eye focuses at infinity. [2 marks]
(b) Calculate the extra power that the lens must produce for the eye to focus on an object that is 30 cm away. [3 marks]

Q2 The eye is designed to receive incoming light and focus it on the retina.
(a) Outline the path of light through the eye, explaining the purpose of the structures it passes through. [4 marks]
(b) Describe the relationship between the structure of the retina and spatial resolution. [2 marks]

The eyes are the window on the soul...

Or so they said in the 16th century. Sadly, that won't get you far with a question about the spectral response of the retina.

Defects of Vision

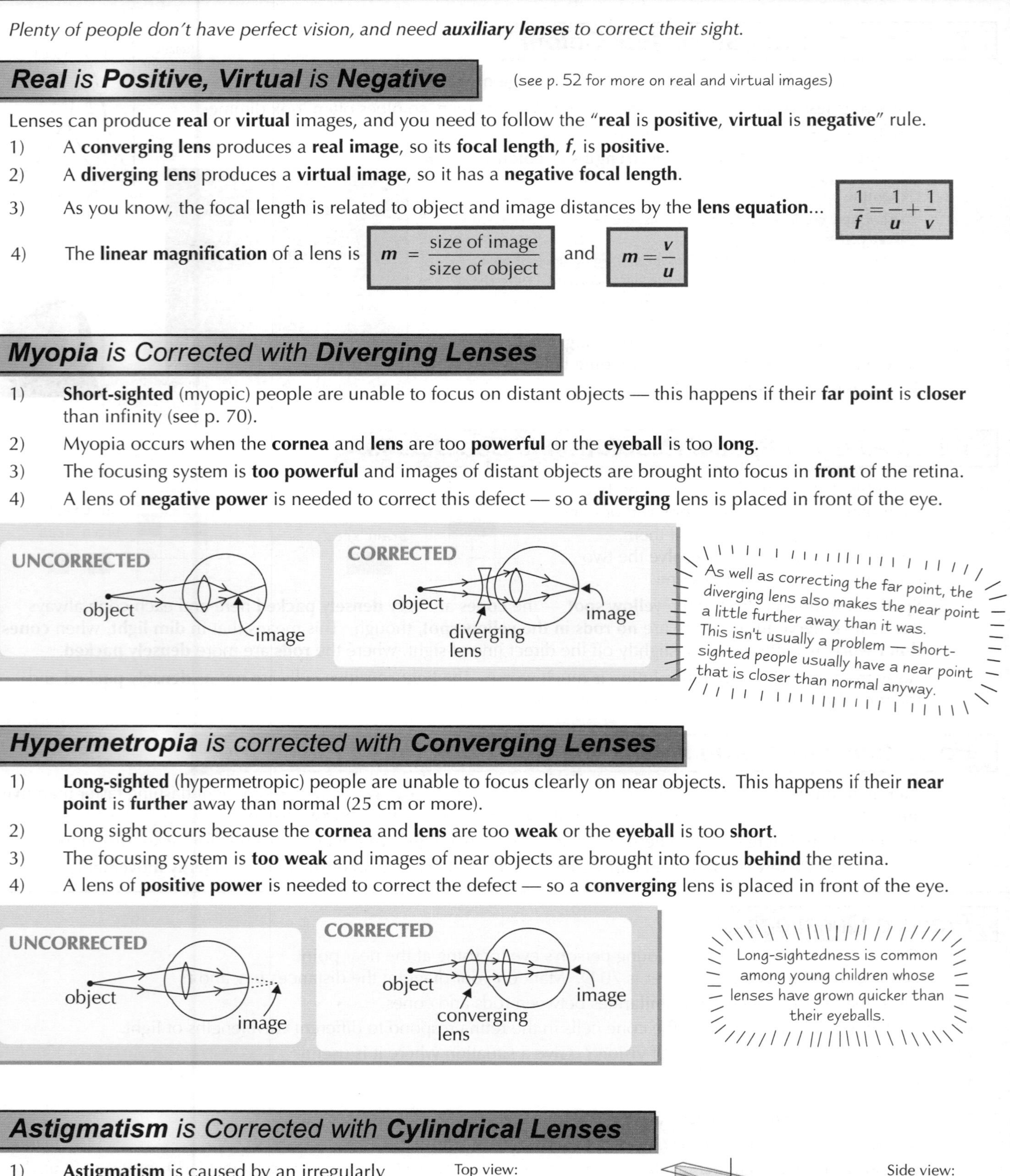

*Plenty of people don't have perfect vision, and need **auxiliary lenses** to correct their sight.*

Real is Positive, Virtual is Negative

(see p. 52 for more on real and virtual images)

Lenses can produce **real** or **virtual** images, and you need to follow the "**real** is **positive**, **virtual** is **negative**" rule.

1) A **converging lens** produces a **real image**, so its **focal length**, ***f***, is **positive**.
2) A **diverging lens** produces a **virtual image**, so it has a **negative focal length**.
3) As you know, the focal length is related to object and image distances by the **lens equation**... $\frac{1}{f} = \frac{1}{u} + \frac{1}{v}$
4) The **linear magnification** of a lens is $m = \frac{\text{size of image}}{\text{size of object}}$ and $m = \frac{v}{u}$

Myopia is Corrected with Diverging Lenses

1) **Short-sighted** (myopic) people are unable to focus on distant objects — this happens if their **far point** is **closer** than infinity (see p. 70).
2) Myopia occurs when the **cornea** and **lens** are too **powerful** or the **eyeball** is too **long**.
3) The focusing system is **too powerful** and images of distant objects are brought into focus in **front** of the retina.
4) A lens of **negative power** is needed to correct this defect — so a **diverging** lens is placed in front of the eye.

UNCORRECTED
object
image

CORRECTED
object
image
diverging lens

As well as correcting the far point, the diverging lens also makes the near point a little further away than it was. This isn't usually a problem — short-sighted people usually have a near point that is closer than normal anyway.

Hypermetropia is corrected with Converging Lenses

1) **Long-sighted** (hypermetropic) people are unable to focus clearly on near objects. This happens if their **near point** is **further** away than normal (25 cm or more).
2) Long sight occurs because the **cornea** and **lens** are too **weak** or the **eyeball** is too **short**.
3) The focusing system is **too weak** and images of near objects are brought into focus **behind** the retina.
4) A lens of **positive power** is needed to correct the defect — so a **converging** lens is placed in front of the eye.

UNCORRECTED
object
image

CORRECTED
object
image
converging lens

Long-sightedness is common among young children whose lenses have grown quicker than their eyeballs.

Astigmatism is Corrected with Cylindrical Lenses

1) **Astigmatism** is caused by an irregularly shaped **cornea** or **lens** which has **different focal lengths** for different **planes**. For instance, when **vertical lines** are in focus, **horizontal** lines might not be.
2) The condition is corrected with **cylindrical lenses**.

Top view:
Rays in horizontal plane are converged

An optician's prescription gives the **angle** of this axis to the **horizontal**

Side view:
Rays in vertical plane are unaffected

Defects of Vision

Choosing a Lens to Correct for Short Sight Depends on the Far Point

1) To correct for **short sight**, a **diverging** lens is chosen which has its **principal focus** at the eye's **faulty far point**.
2) The **principal focus** is the point that rays from a distant object **appear** to have come from (see p.52).
3) The lens must have a **negative focal length** which is the same as the **distance to the eye's far point**. This means that objects at **infinity**, which were out of focus, now seem to be in focus at the far point.

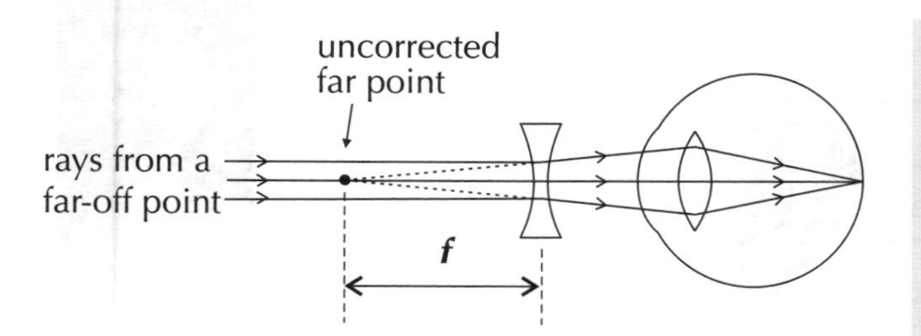

Example

Ben is short-sighted. His far point is 5 m.
Calculate the power of lens he needs to correct his vision.

Focal length, f = far point = −5 m

Power needed = $\frac{1}{f}$ = **−0.2 D**

The power's always **negative** to correct for **short** sight.

Calculations Involving Long Sight Use the Lens Equation

1) People with these conditions have a near point which is too far away. An 'acceptable' near point is 25 cm.
2) A **converging lens** is used to produce a **virtual image** of objects 0.25 m away **at the eye's near point.** This means that close objects, which were out of focus, now seem to be in focus at the near point.
3) You can work out the **focal length**, and hence the **power** of lens needed, using the **lens equation** $\frac{1}{f} = \frac{1}{u} + \frac{1}{v}$

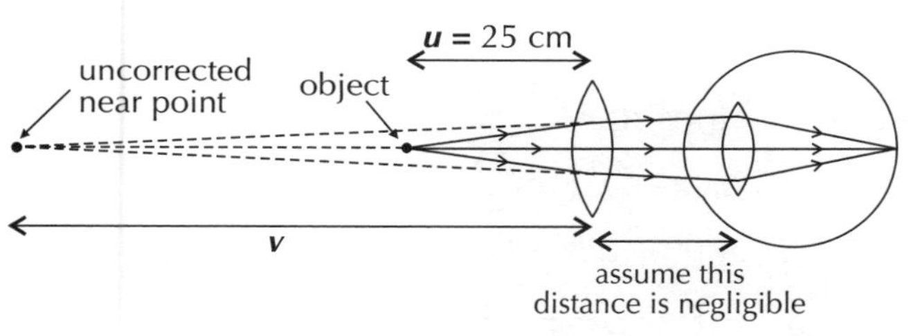

Example

Mavis can't read her book — her near point is 5 m.
What power of lens does she need?

u = 0.25 m, v = −5 m (real is positive, virtual is negative)

$$\frac{1}{f} = \frac{1}{0.25} - \frac{1}{5}$$

$$\frac{1}{f} = 3.8 \text{ so power} = +3.8 \text{ D}$$

The power's always **positive** to correct for **long** sight.

Practice Questions

Q1 Define the terms myopia, hypermetropia and astigmatism.

Q2 What type of auxiliary lenses are used to correct these conditions?

Q3 Define the terms near point and far point.

Exam Questions

Q1 A man with short sight has a far point of 4 m.
Calculate the power of auxiliary lens needed to correct his far point. [3 marks]

Q2 A girl has a near point of 2 m.
Calculate the power of lens required to correct her near point to 25 cm. [3 marks]

Q3 Claire suffers from astigmatism.
(a) State the type of lenses that are used to correct astigmatism. [1 mark]
(b) Draw a diagram of a lens that would converge rays in the vertical plane but not in the horizontal plane. [2 marks]

You can't fly fighter planes if you wear glasses...

There's a hidden bonus to having dodgy eyes — in the exam, you can take your specs off (discreetly) and have a look at the lenses to remind yourself what type is needed to correct short sight, long sight, or whatever it is you have. Cunning.

Physics of the Ear

Ears are pretty amazing — they convert sound into electrical energy, using some tiny bones and lots of even tinier hairs.

The **Intensity** of **Sound** is **Power** per **Unit Area**

The **intensity** of a sound wave is defined as the amount of sound **energy** that passes **per second per unit area** (perpendicular to the direction of the wave). That's **power per unit area**.

1) If the sound energy arriving at the ear per second is ***P***, then the intensity of the sound is:

$$I = \frac{P}{A}$$

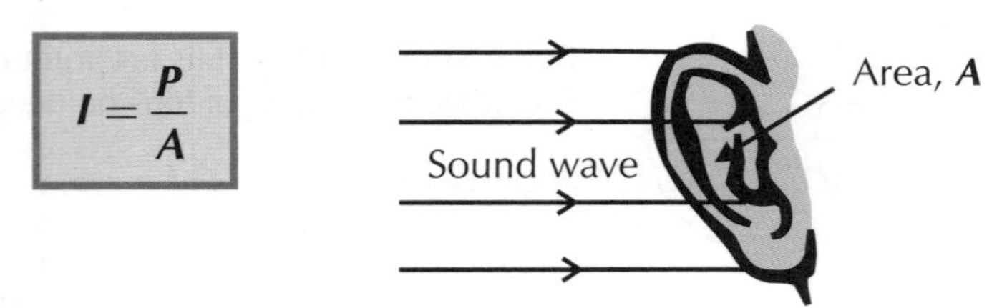

Felicity thought her waves were sound.

2) The SI unit of intensity is Wm^{-2}, but you'll often see decibels used instead (see p. 76).
3) For any wave, **intensity ∝ amplitude²** — so doubling the amplitude will result in four times the intensity.
4) Intensity is related to the **loudness** of sound (see p. 76).

The **Ear** has **Three Main Sections**

The ear consists of three sections: the **outer** ear (**pinna** and **auditory canal**), the **middle ear** (**ossicles** and **Eustachian tube**) and the **inner ear** (**semicircular canals**, **cochlea** and **auditory nerve**).

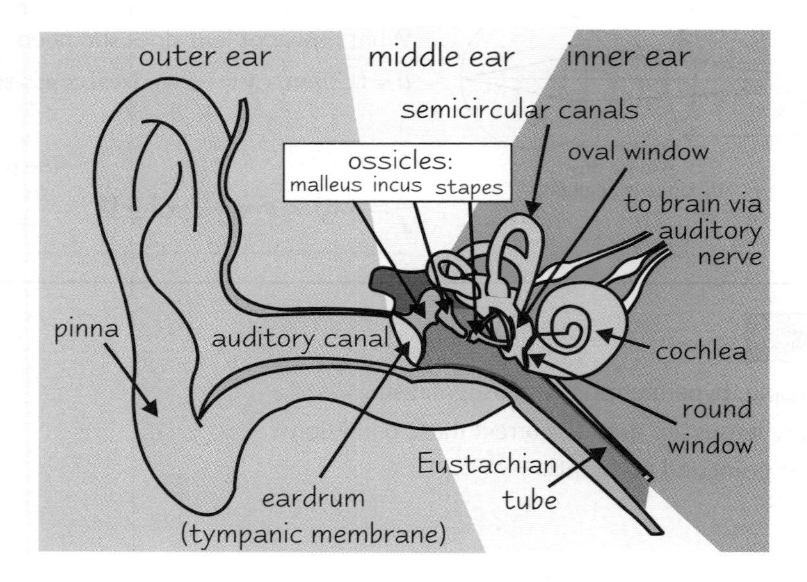

1) The **tympanic membrane** (eardrum) separates the **outer** and **middle** ears.
2) Although separated, the **outer** and **middle** ears both contain **air** at **atmospheric pressure**, apart from slight pressure variations due to sound waves. This pressure is maintained by **yawning** and **swallowing** — the middle ear is opened up to the outside via the **Eustachian tube** (which is connected to the throat).
3) The **oval** and **round windows** separate the **middle** and **inner** ears.
4) The **inner ear** is filled with fluid called **perilymph** (or **endolymph** in the **cochlear duct**). This fluid allows **vibrations** to pass to the basilar membrane in the **cochlea**.
5) The **semicircular canals** are involved with **maintaining balance**.

Physics of the Ear

The **Ear** acts as a **Transducer,** converting **Sound Energy**...

1) The **pinna** (external ear) acts like a funnel, channelling sound waves into the auditory canal — this **concentrates** the energy onto a **smaller area**, which increases the **intensity**.
2) The sound waves consist of **variations** in **air pressure**, which make the **tympanic membrane** (eardrum) **vibrate**.
3) The tympanic membrane is connected to the **malleus** — one of the **three tiny bones** (**ossicles**) in the middle ear. The malleus then passes the **vibrations** of the eardrum on to the **incus** and the **stapes** (which is connected to the **oval window**).
4) As well as **transmitting vibrations**, the **ossicles** have **two** other functions — **amplifying** the sound signal and **reducing** the **energy reflected back** from the inner ear.
5) The **oval window** has a much **smaller area** than the **tympanic membrane**. Together with the **increased force** produced by the ossicles, this results in **greater pressure variations** at the oval window.
6) The **oval window** transmits vibrations to the **fluid** in the **inner ear**.

...into **Electrical Energy**

1) Pressure waves in the fluid of the **cochlea** make the **basilar membrane** vibrate. Different regions of this membrane have different **natural frequencies**, from 20 000 Hz near the middle ear to 20 Hz at the other end.
2) When a sound wave of a particular **frequency** enters the inner ear, one part of the basilar membrane **resonates** and so vibrates with a **large amplitude**.
3) **Hair cells** attached to the basilar membrane trigger **nerve impulses** at this point of greatest vibration.
4) These **electrical impulses** are sent, via the **auditory nerve**, to the **brain**, where they are interpreted as **sounds**.

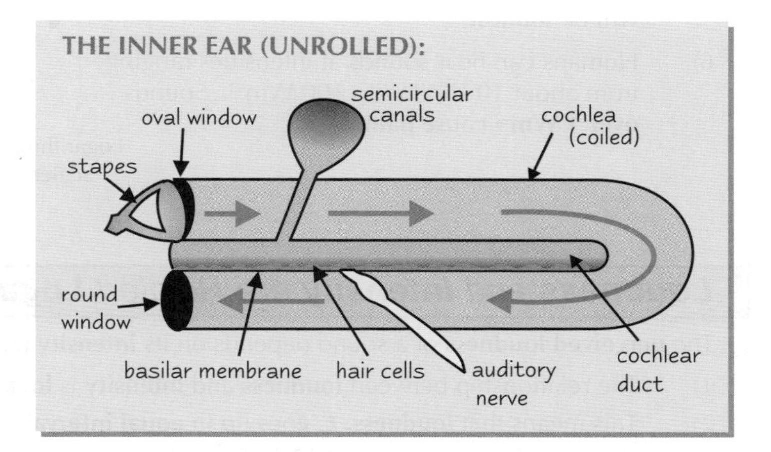

Practice Questions

Q1 What is meant by the 'intensity' of sound? What is the formula for intensity?

Q2 Sketch a diagram of the ear, labelling the structures within it.

Q3 Describe the function of the ossicles.

Q4 Explain why the relative size of the oval window and tympanic membrane is important.

Exam Question

Q1 The ear is designed to transduce sound energy into electrical energy.

(a) State the function of the pinna. [1 mark]

(b) Describe how sound energy is transmitted through the middle ear. [3 marks]

(c) The surface area of the tympanic membrane is around 14 times the area of the oval window. Show that this increases the amplitude of vibrations in the ear by a factor of approximately 3.74. [3 marks]

(d) Describe how pressure waves in the cochlea are converted to electrical impulses. [2 marks]

(e) Explain how the ear is able to encode the frequency of a sound in the information sent to the brain. [2 marks]

Ears are like essays — they have a beginning, middle and end...

Or outer, middle and inner, if we're being technical. Learn what vibrates where, and you'll be fine.

Intensity and Loudness

*The ear's sensitivity depends on the **frequency** and **intensity** of sounds, and deteriorates as you get older.*

Humans can Hear a Limited Range of Frequencies

1) Young people can hear frequencies ranging from about **20 Hz** (low pitch) up to **20 000 Hz** (high pitch). As you get older, the upper limit decreases.
2) Our ability to **discriminate between frequencies** depends on how **high** that frequency is. For example, between 60 and 1000 Hz, you can hear frequencies 3 Hz apart as **different pitches**. At **higher** frequencies, a **greater difference** is needed for frequencies to be distinguished. Above 10 000 Hz, pitch can hardly be discriminated at all.
3) The **loudness** of sound you hear depends on the **intensity** and **frequency** of the sound waves.
4) The **weakest intensity** you can hear — the **threshold of hearing**, I_0 — depends on the **frequency** of the sound wave.
5) The ear is **most sensitive** at around **3000 Hz**. For any given intensity, sounds of this frequency will be **loudest**.
6) Humans can hear sounds at intensities ranging from about 10^{-12} Wm^{-2} to 100 Wm^{-2}. Sounds **over 1 Wm^{-2}** cause **pain**.

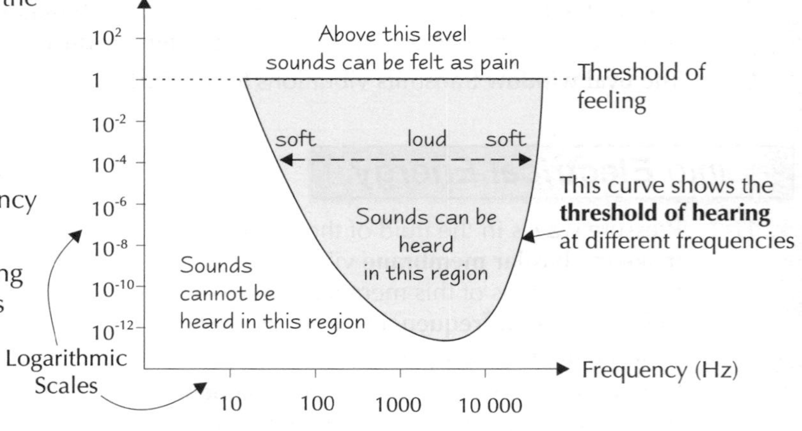

Loudness and Intensity are Related Logarithmically

The **perceived loudness** of a sound depends on its **intensity** (and its frequency — see above).

1) The relationship between **loudness** and **intensity** is **logarithmic**.
2) This means that **loudness**, L, goes up in **equal intervals** if **intensity**, I, increases by a **constant factor** (provided the frequency of the sound doesn't change).
3) E.g. if you **double** the intensity, **double it again** and so on, the **loudness** keeps going up in **equal steps**.

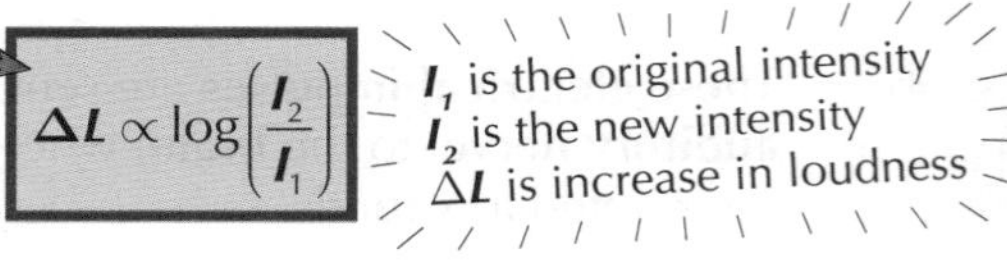

The Decibel Scale is used for Measuring Intensity Level

1) You can often measure loudness using a **decibel meter**. The **decibel scale** is a **logarithmic scale** which actually measures **intensity level**.
2) The **intensity level**, IL, of a sound of intensity I is defined as

$$IL = 10\log\left(\frac{I}{I_0}\right)$$

I = intensity
I_0 = threshold of hearing

3) I_0 is the **threshold of hearing** (the **lowest intensity** of sound that can be heard) at a frequency of **1000 Hz**.
4) The value of I_0 is **1×10^{-12} Wm^{-2}**.
5) The units of IL are **decibels** (dB). Intensity level can be given in **bels** — one decibel is a tenth of a bel — but decibels are usually a more convenient size.

The dBA Scale is an Adjusted Decibel Scale

1) The **perceived loudness** of a sound depends on its **frequency** as well as its intensity. Two different frequencies with the **same loudness** will have **different intensity levels** on the dB scale.
2) The **dBA** scale is an **adjusted decibel scale** which is designed to take into account the **ear's response** to **different frequencies**.
3) On the **dBA scale**, sounds of the **same intensity level** have the **same loudness** for the average human ear.

Intensity and Loudness

You can Generate *Curves of Equal Loudness*

1) Start by generating a **control frequency** of **1000 Hz** at a particular **intensity level**.
2) Generate another sound at a different frequency. Vary the volume of this sound until it appears to have the **same loudness** as the 1000 Hz frequency. Measure the **intensity level** at this volume.
3) Repeat this for several different frequencies, and plot the resulting curve on a graph.
4) Change the **intensity level** of the **control frequency** and repeat steps two and three.
5) If you measure **intensity level** in **decibels**, then the **loudness** of the sound is given in **phons**.

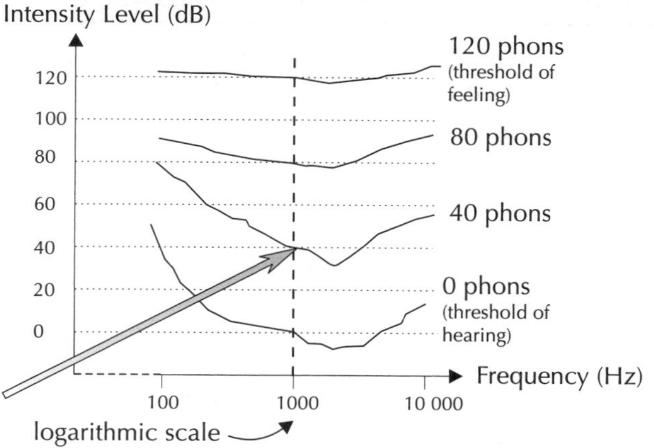

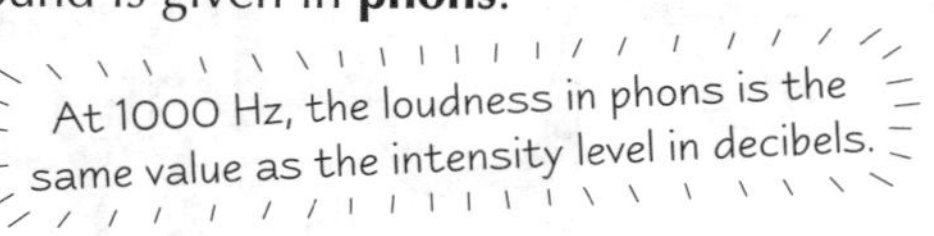

Hearing Deteriorates with *Age* and *Exposure* to *Excessive Noise*

1) As you get **older**, your hearing deteriorates **generally**, but **higher frequencies** are affected **most**.
2) Your ears can be damaged by **excessive noise**. This results in general hearing loss, but frequencies around **4000 Hz** are usually worst affected.
3) People who've worked with very **noisy machinery** have most hearing loss at the **particular frequencies** of the noise causing the damage.
4) **Equal loudness curves** can show hearing loss.
5) For a person with hearing loss, **higher intensity levels** are needed for the **same loudness**, when compared to a normal ear. A **peak** in the curve shows damage at a **particular** range of **frequencies**.

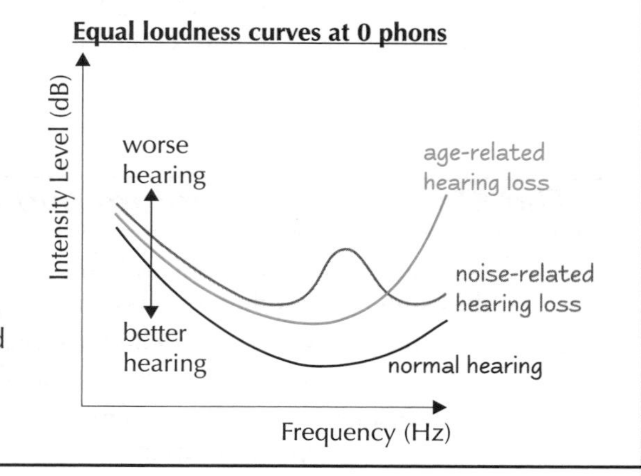

Practice Questions

Q1 Define the threshold of hearing and sketch a graph that shows how it depends on frequency.

Q2 How are curves of equal loudness generated?

Q3 What is the dB scale? How is the dBA scale different?

Exam Questions

Q1 A small siren, which can be regarded as a point source, emits sound waves at a frequency of 3000 Hz. The intensity of the sound is 0.94 Wm^{-2} at a distance of 10 m.

(a) State the accepted value of the threshold of hearing at 1000 Hz, I_0, in Wm^{-2}. [1 mark]

(b) Calculate the intensity level of the sound of the siren. [2 marks]

(c) Explain why the siren uses a frequency of 3000 Hz. [1 mark]

Q2 The diagram shows an equal loudness curve for a person suffering hearing loss and a person with normal hearing. The patient believes his hearing may have been damaged by working with noisy machinery. Does his equal loudness curve support this? Explain your answer. [3 marks]

Intensity Level (dB)
patient with hearing loss
normal hearing
Frequency (Hz)

Saved by the decibel....

It's medical fact that prolonged loud noise damages your hearing, so you should really demand ear protection before you agree to do the housework — some vacuum cleaners are louder than 85 dBA — the 'safe' limit for regular exposure.

Physics of the Heart

There's a bit of biochemistry on this page — my, my, aren't you lucky people...

The Heart is a Double Pump

1) The heart is a **large muscle**. It acts as a **double pump**, with the **left**-hand side pumping blood from the **lungs** to the **rest of the body** and the **right**-hand side pumping blood from the **body** back to the **lungs**.
2) Traditionally, a diagram of the heart is drawn as though you're looking at it **from the front**, so the **right**-hand side of the heart is drawn on the **left**-hand side of the **diagram** and vice versa (just to confuse you).
3) Each side of the heart has **two chambers** — an **atrium** and a **ventricle** — separated by a **valve**.
4) **Blood** enters the **atria** from the veins, then the atria **contract**, squeezing blood into the **ventricles**. The **ventricles** then **contract**, squeezing the blood **out** of the heart into the **arteries**. The **valves** are there so that the blood doesn't go back into the atria when the ventricles contract.

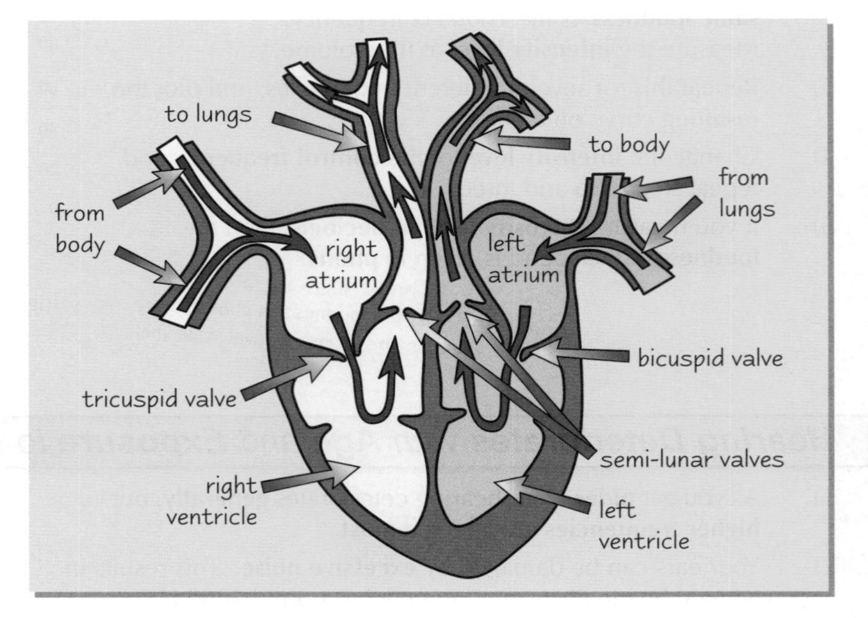

The Movement of Sodium and Potassium Ions Generates Electrical Signals

1) The movement of **sodium** and **potassium ions** across **cell membranes** in the heart is controlled by **diffusion** and a **sodium-potassium pump**.
2) The **pump** uses a series of **chemical reactions** to move **sodium ions** (Na^+) through the membrane and out of the cell, and to move **potassium ions** (K^+) into the cell.
3) By **diffusion**, particles tend to move from **high** to **low concentrations** (along a **concentration gradient**). So, as the pump builds up the ion concentrations, the ions try to **diffuse back** in the **opposite** direction.
4) So far, so good — but here's where it gets a bit more complicated.
5) For most of the time, the membrane **only lets K^+ ions** diffuse through, and **not** the Na^+ ions. That means there's a **net flow** of **positive charge** **out of** the cell and a **voltage** builds up until equilibrium is reached.
6) By convention, the potential on the **outside** of the membrane (where there's lots of sodium) is taken to be **0 V**, so the membrane potential on the inside is **negative** (–80 mV for heart muscle). You've now got a **polarised** membrane.

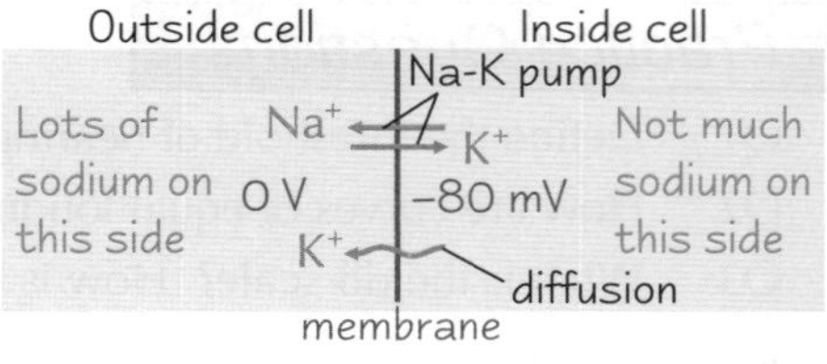

7) If the membrane is **stimulated** by an **electrical signal** it suddenly becomes **permeable** to **Na^+** as well. The sodium ions **rush** across the membrane towards the **negative side** (inside the cell), pushed by both the **Na^+ concentration gradient** and **electrostatic forces**.
8) This first **depolarises** the membrane, then **charges it up** the **other way** (called reverse-polarisation), reaching a **positive potential** of about 40 mV. This makes the heart muscle **CONTRACT**.
9) The membrane then becomes **impermeable** to **Na^+** ions again, but **very permeable** to **K^+** ions. The K^+ ions **diffuse** very quickly, repolarising the membrane, and the heart muscle **RELAXES**.
10) The Na-K pump then slowly restores the potential back to its **equilibrium polarised state**.

The sudden flip in potential is called the **action potential**. When this happens at one part of a membrane, it triggers the part next to it to do the same and so an electrical signal passes down the membrane.

Physics of the Heart

The Heart's **Pacemaker** is the **Sinoatrial Node**

1) A group of specialised cells at the **sinoatrial (SA) node** (in the wall of the right atrium) produce **electrical signals** that pulse about **70 times a minute**.
2) These signals spread through the **atria** and make them **contract** via the **action potential** (see previous page).
3) The signals then pass to the **atrioventricular (AV) node**, which **delays** the pulse for about **0.1 seconds** before passing it on to the **ventricles**.
4) The ventricles **contract** and the process repeats.

The Heart can be **Monitored** by an **Electrocardiograph (ECG)**

1) The **potential difference** between the **polarised** and **depolarised** heart cells produces a **weak electrical signal** at the surface of the body. This is plotted against time to give an **electrocardiogram (ECG)**, which can provide useful information about the **condition** of the heart.
2) A **normal** ECG, covering a **single heartbeat**, has **three** separate parts: a **P** wave, a **QRS** wave and a **T** wave.
3) The **P wave** corresponds to the **depolarisation** and **contraction** of the **atria**.
4) The **QRS wave** (about 0.2 seconds later) corresponds to the **depolarisation** and **contraction** of the **ventricles**. This completely swamps the trace produced by the repolarisation and relaxation of the atria.
5) Finally, the **T wave** (another 0.2 seconds later) corresponds to the **repolarisation** and **relaxation** of the **ventricles**.

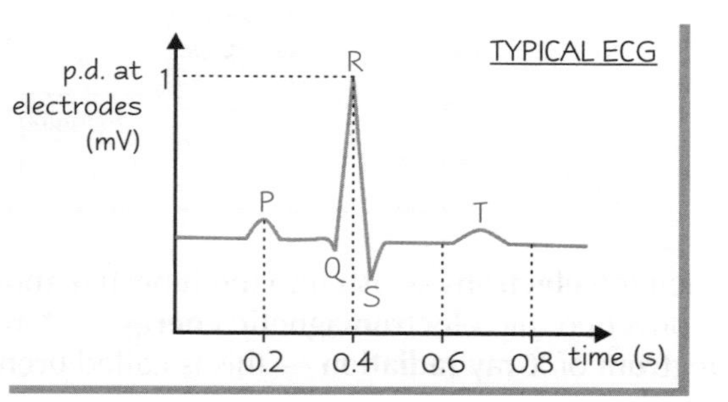

6) There are **12** standard ways of placing electrodes on the body to obtain an **ECG**, each producing a slightly different waveform. In all cases the signal is **heavily attenuated** (absorbed and weakened) by the body and needs to be amplified by a high impedance **amplifier**.
7) **Electrodes** are placed on the **chest** and the **limbs** where the arteries are close to the surface. The **right leg** is **never** used since it is **too far away** from the heart.
8) In order to get a **good electrical contact**, **hairs** and **dead skin** cells are removed and a **conductive gel** is used.

Practice Questions

Q1 Describe the basic structure of the heart, and the passage of blood through the heart, lungs and the body.

Q2 Describe how a membrane in heart muscle is initially polarised and what happens when it is stimulated.

Q3 Sketch a typical ECG trace and indicate the main features.

Exam Questions

Q1 A patient with a suspected heart condition is given an ECG.

(a) Describe how the patient's skin is prepared to ensure a good electrical contact with the electrodes. [2 marks]

(b) The patient's ECG shows a pause between the P and QRS waves.
Describe how the action of the heart corresponds to these waves. [3 marks]

Q2 Explain the processes of depolarisation and repolarisation in the formation of an action potential.
The quality of your written answer will be assessed in this question. [6 marks]

Q3 Outline how electrical signals are involved in the function of the heart. [4 marks]

Be still my beating sinoatrial node...

If you rely on the cast of ER to get your heart beating faster, console yourself that it's all very educational. Listen out for the machine that goes 'bip, bip, bip', and look for the P waves, QRS waves and T waves on the screen. If there aren't enough waves, the brave docs have to start shouting 'clear' and waving defibrillators around.

X-Ray Production

X-ray imaging is one kind of non-invasive diagnostic technique — these techniques let doctors see what's going on (or going wrong) inside your body, without having to open you up and have a look.

X-rays are Produced by Bombarding Tungsten with High Energy Electrons

1) In an X-ray tube, **electrons** are emitted from a **heated filament** and **accelerated** through a high **potential difference** (the **tube voltage**) towards a **tungsten anode**.

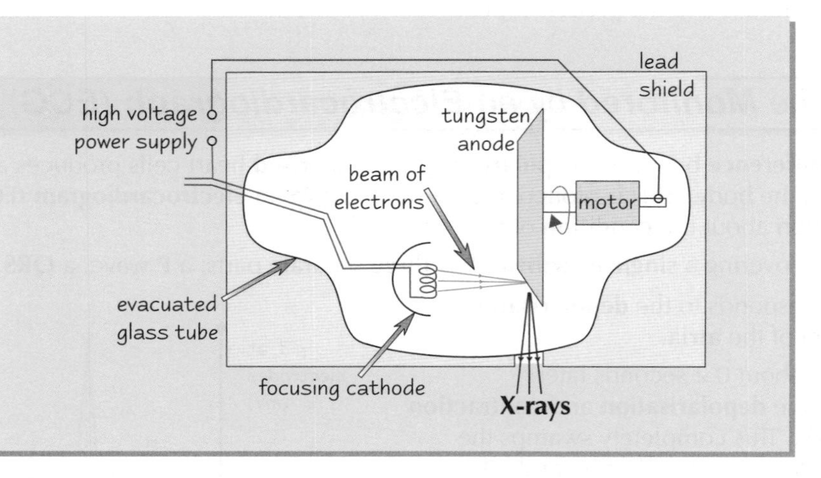

2) When the **electrons** smash into the **tungsten anode**, they **decelerate** and some of their **kinetic energy** is converted into **electromagnetic energy**, as **X-ray photons**. The tungsten anode emits a **continuous spectrum** of **X-ray radiation** — this is called **bremsstrahlung** ('braking radiation').

3) 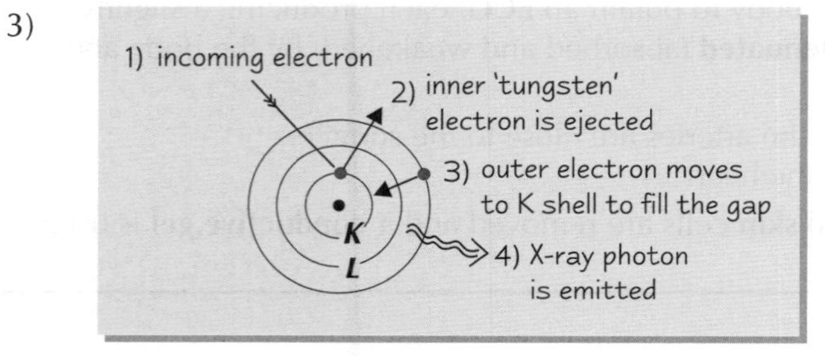

X-rays are also produced when beam electrons **knock out** other electrons from the **inner shells** of the **tungsten atoms**.

Electrons in the atoms' **outer shells** move into the **vacancies** in the **lower energy levels**, and **release energy** in the form of **X-ray photons**.

4) This process results in **line spectra** superimposed on a **continuous spectrum**.

Relative Intensity
continuous radiation (bremsstrahlung)
K lines
L lines
E_{max}
Photon Energy (= hf)

The K and L lines show which shell the electrons are filling vacancies in to cause the release of energy.

Brian's glasses gave a whole new meaning to 'undressing with your eyes'.

5) Only about **1%** of the electrons' **kinetic energy** is converted into **X-rays**. The rest is converted into **heat**, so, to avoid overheating, the tungsten anode is **rotated** at about 3000 rpm. It's also **mounted** on **copper** — this **conducts** the heat away effectively.

X-Ray Production

Beam Intensity and **Photon Energy** can be Varied

The **intensity** of the X-ray beam is the **energy per second per unit area** passing through a surface (at right angles). There are two ways to increase the **intensity** of the X-ray beam:

1) Increase the **tube voltage**. This gives the electrons **more kinetic energy**. Higher energy electrons can **knock out** electrons from shells **deeper** within the tungsten atoms — giving more 'spikes' on the graphs. Individual **X-ray photons** also have **higher maximum energies**.

   Intensity is approximately **proportional** to **voltage squared**.

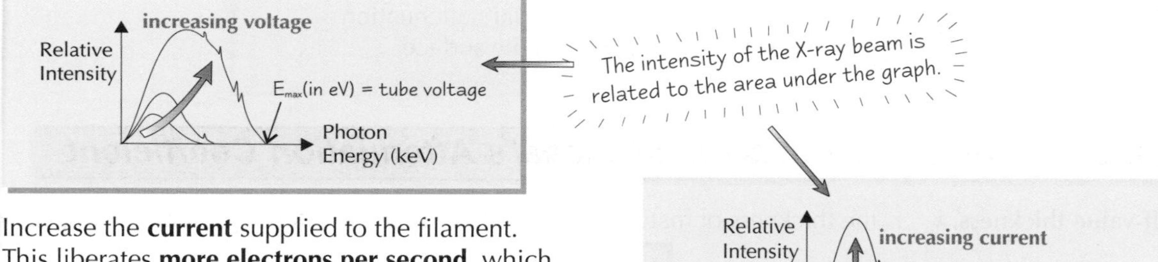

2) Increase the **current** supplied to the filament. This liberates **more electrons per second**, which then produce **more X-ray photons per second**. Individual **photons** have the **same energy** as before.

   Intensity is approximately **proportional** to **current**.

Radiographers try to Produce a **Sharp Image** and **Minimise** the **Radiation Dose**

Medical X-rays are a compromise between producing really sharp, clear images, whilst keeping the amount of radiation the patient is exposed to as low as possible. To do this, radiographers:

1) Put the **detection plate close** to the patient and the **X-ray tube far** away from the patient.
2) Make sure the patient **keeps still** — if they move around, the image will be blurred.
3) Put a **lead grid** between the patient and film to **stop** scattered radiation '**fogging**' the film and **reducing contrast**.
4) Use an **intensifying screen** next to the film surface. This consists of crystals that **fluoresce** — they **absorb X-rays** and re-emit the energy as **visible light**, which helps to develop the photograph quickly. A shorter exposure time is needed, keeping the patient's radiation dose lower.

Practice Questions

Q1 Draw a diagram of an X-ray tube and explain how a typical X-ray spectrum is produced.

Q2 What is meant by the 'intensity'of an X-ray beam?

Q3 Give two ways in which the intensity of an X-ray beam can be increased.

Q4 What measures can be taken to produce a high quality X-ray image while reducing the patient's radiation dose?

Exam Question

(Charge on an electron (e) = -1.6×10^{-19} J. Mass of an electron (m_e) = 9.11×10^{-31} kg.)

Q1 An X-ray tube is connected to a potential difference of 30 kV.

(a) Sketch a graph of relative intensity against photon energy (in eV) for the resulting X-ray spectrum, and indicate its main features. [3 marks]

(b) Show how the graph in (a) would change if the tube voltage were increased. [2 marks]

(c) Calculate the velocity of the electrons arriving at the anode. [4 marks]

Situation vacant — electron needed for low energy position...

I have a question — why, when something could have a nice, simple name like 'braking radiation', do scientists insist on giving it a much fancier one? 'Bremsstrahlung' just sounds baffling — well, unless you speak German of course.

X-Ray and MRI Imaging

So, you know how X-rays are produced and what the radiographer does — but why, I hear you cry, do some bits of you (i.e. your bones) show up nicely in an X-ray image, while others fade into the background? Attenuation, that's why...

X-Rays are *Attenuated* when they *Pass Through Matter*

When X-rays pass through matter (e.g. a patient's body), they are **absorbed** and **scattered**. The intensity of the X-ray beam **decreases** (attenuates) **exponentially** with the **distance from** the **surface**, according to the material's **attentuation coefficient**.

$$I = I_0 e^{-\mu x}$$

where I is the intensity of the X-ray beam, I_0 is the initial intensity, μ is the material's attenuation coefficient and x is the distance from the surface.

Half-value Thickness Depends on a Material's *Attenuation Coefficient*

1) **Half-value thickness**, $x_{\frac{1}{2}}$, is the thickness of material required to **reduce** the **intensity** to **half** its **original value**.

   This depends on the **attenuation coefficient** of the material, and is given by:

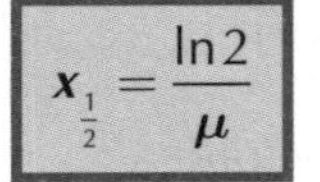

$$x_{\frac{1}{2}} = \frac{\ln 2}{\mu}$$

2) The **mass attenuation coefficient**, μ_m, for a material of density ρ is given by: $$\mu_m = \frac{\mu}{\rho}$$

X-rays are *Absorbed* More by *Bone* than *Soft Tissue*

1) X-rays are **attenuated** by **absorption** and **scattering**. How much **energy is absorbed** by a **material** depends on its **atomic number**.
2) So tissues containing atoms with **different atomic numbers** (e.g. **soft tissue** and **bone**) will **contrast** in the X-ray image.
3) If the tissues in the region of interest have similar attenuation coefficients then artificial **contrast media** can be used — e.g. **barium meal**.
4) **Barium** has a **high atomic number**, so it shows up clearly in X-ray images and can be followed as it moves along the patient's digestive tract.

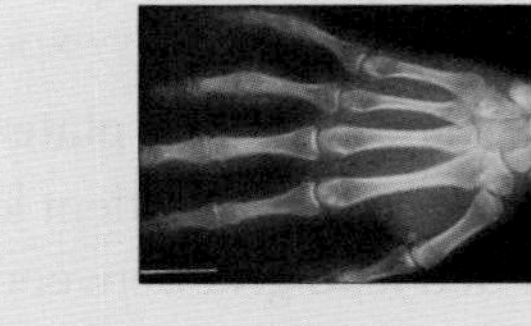

Bones show up brightly in X-ray images because they absorb more X-rays than the surrounding soft tissue.

CT Scans use *X-rays* to Produce *High-Quality Images*

1) **Computed tomography** (CT) scans produce an image of a **two-dimensional slice** through the body.
2) A narrow **X-ray beam** consisting of a single wavelength (**monochromatic**) **rotates** around the body and is picked up by thousands of **detectors**. The detectors feed the signal to a **computer**.
3) The computer works out how much attenuation has been caused by each part of the body and produces a very **high quality** image.
4) However, the machines are **expensive** and the scans involve a **high radiation dose** for the patient.

Fluoroscopy is used to Create *Moving Images*

1) **Moving images** can be created using a **fluorescent screen** and an **image intensifier**. This is useful for **imaging organs** as they **work**.
2) **X-rays** pass through the patient and hit the **fluorescent screen**, which **emits light**.
3) The **light** causes **electrons** to be emitted from the **photocathode**.
4) The **electrons** travel through the **glass tube** towards the **fluorescent viewing screen**, where they form an image. Electrodes in the glass tube **focus** the **image** onto the screen.

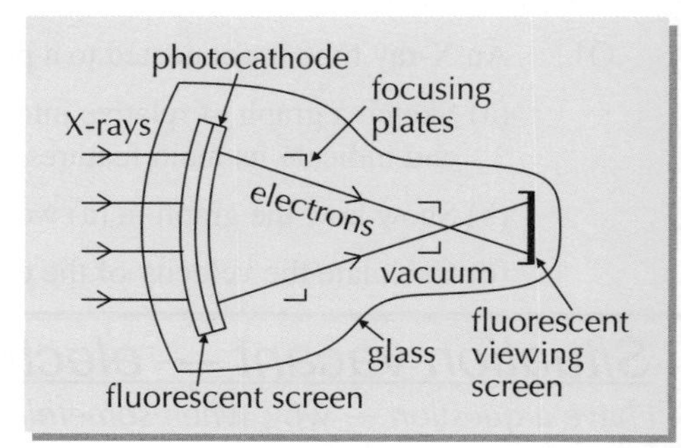

X-Ray and MRI Imaging

Magnetic Resonance Imaging, or MRI to you and me, is another form of non-invasive diagnostic imaging — enjoy.

Magnetic Resonance can be used to Create Images

1) The patient lies in the centre of a huge **superconducting magnet** that produces a **uniform magnetic field**. The magnet needs to be **cooled** by **liquid helium** — this is partly why the scanner is so **expensive**.
2) Radio frequency **coils** are used to transmit **radio waves**, which **excite hydrogen nuclei** in the patient's body.
3) When the radio waves are switched off, the hydrogen nuclei relax and emit electromagnetic energy — this is the **MRI signal** (more details below). The radio frequency coils **receive the signal** and send it to a **computer**.
4) The computer **measures** various quantities of the MRI signal — amplitude, frequency, phase — and **analyses** them to generate an **image** of a **cross-section** through the body.

Contrast can be Controlled by Varying the Pulses of Radio Waves

1) Radio waves are applied in **pulses**. Each short pulse **excites** the hydrogen nuclei and then allows them to **relax** and emit a signal. The response of **different tissue types** can be enhanced by varying the **time between pulses**.
2) Tissues consisting of **large molecules** such as **fat** are best imaged using **rapidly repeated pulses**. This technique is used to image the internal **structure** of the body.
3) Allowing **more time** between pulses enhances the response of **watery** substances. This is used for **diseased** areas.

MRI has Advantages and Disadvantages

ADVANTAGES:

1) There are **no** known **side effects**.
2) An image can be made for any slice in any **orientation** of the body.
3) High quality images can be obtained for **soft tissue** such as the **brain**.
4) **Contrast** can be **weighted** in order to investigate different situations.

DISADVANTAGES:

1) The imaging of **bones** is very **poor**.
2) Some people suffer from **claustrophobia** in the scanner.
3) Scans can be **noisy** and take a **long time**.
4) MRI can't be used on people with **pacemakers** or some **metal implants** — the strong magnetic fields would be very harmful.
5) Scanners **cost millions** of pounds.

Practice Questions

Q1 Outline why bone shows up in X-ray images.

Q2 What are 'artificial contrast media' used for? Give an example.

Q3 Give two advantages of MRI compared to X-ray imaging.

Exam Questions

Q1 The half-value thickness for aluminium is 3 mm for 30 keV X-ray photons.

(a) Explain what is meant by the term 'half-value thickness'. [1 mark]

(b) Determine the thickness of aluminium needed to reduce the intensity of a homogeneous beam of X-rays at 30 keV to 1% of its initial value. [4 marks]

Q2 Outline how an MRI scanner is used to produce an image of a section of a patient's body.
The quality of your written answer will be assessed in this question. [6 marks]

Q3 Discuss the advantages and disadvantages of MRI scanning as a medical imaging technique. [6 marks]

I've got attenuation coefficient disorder — I get bored really easily...

OK, so it hasn't been the easiest of pages. But at least now you know why people sit in vats of baked beans to raise money for their local hospital to buy an MRI scanner. Though perhaps you need A2 Psychology to understand the beans part.

Ultrasound Imaging

Ultrasound is a 'sound' with higher frequencies than we can hear.

Ultrasound has a Higher Frequency than Humans can Hear

1) Ultrasound waves are **longitudinal** waves with **higher frequencies** than humans can hear (>20 000 Hz).
2) For **medical** purposes, frequencies are usually from **1** to **15 MHz**.
3) When an ultrasound wave meets a **boundary** between two **different materials**, some of it is **reflected** and some of it passes through (undergoing **refraction** if the **angle of incidence** is **not 90°**).
4) The **reflected waves** are detected by the **ultrasound scanner** and are used to **generate an image**.

The Amount of Reflection depends on the Change in Acoustic Impedance

1) The **acoustic impedance**, Z, of a medium is defined as: $Z = \rho c$
Z has units of $kgm^{-2}s^{-1}$.

ρ = density of the material, in kgm^{-3}
c = speed of sound in the medium, in ms^{-1}

2) Say an ultrasound wave travels through a material with an impedance Z_1. It hits the boundary between this material and another with an impedance Z_2. The incident wave has an intensity of I_i.
3) If the two materials have a **large difference** in **impedance**, then **most** of the energy is **reflected** (the intensity of the reflected wave I_r will be high). If the impedance of the two materials is the **same** then there is **no reflection**.
4) The **fraction** of wave **intensity** that is reflected is called the **intensity reflection coefficient**, α.

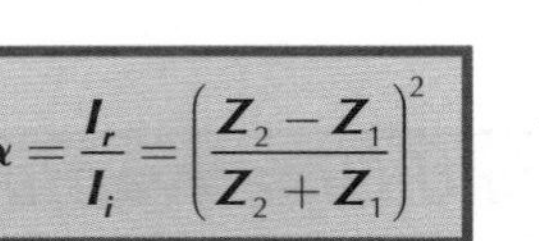

$$\alpha = \frac{I_r}{I_i} = \left(\frac{Z_2 - Z_1}{Z_2 + Z_1}\right)^2$$

You don't need to learn this equation. Just practise using it.

There are Advantages and Disadvantages to Ultrasound Imaging

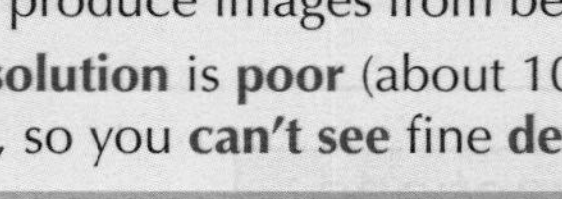

ADVANTAGES:

1) There are **no** known **hazards** — in particular, **no** exposure to **ionising radiation**.
2) It's good for imaging **soft tissues**, since you can obtain **real-time** images — X-ray fluoroscopy can achieve this, but involves a huge dose of radiation.
3) Ultrasound devices are relatively **cheap** and **portable**.

DISADVANTAGES:

1) Ultrasound **doesn't penetrate bone** — so it **can't** be used to **detect fractures** or examine the **brain**.
2) Ultrasound **cannot** pass through **air spaces** in the body (due to the **mismatch** in **impedance**) — so it can't produce images from behind the lungs.
3) The **resolution** is **poor** (about 10 times worse than X-rays), so you **can't see** fine **detail**.

Ultrasound Images are Produced Using the Piezoelectric Effect

1) **Piezoelectric crystals** produce a **potential difference** when they are **deformed** (squashed or stretched) — the rearrangement in structure displaces the **centres of symmetry** of their electric **charges**.

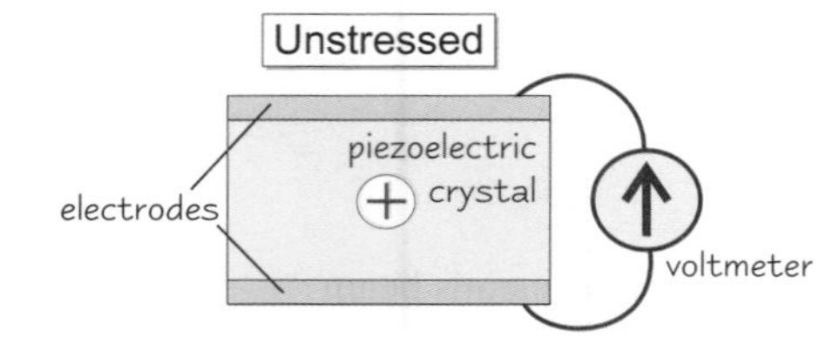

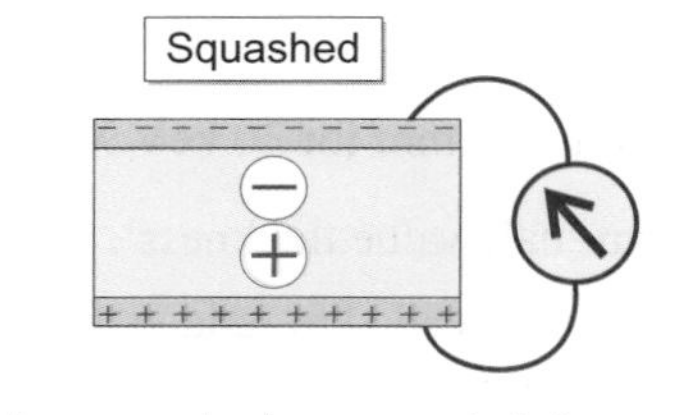

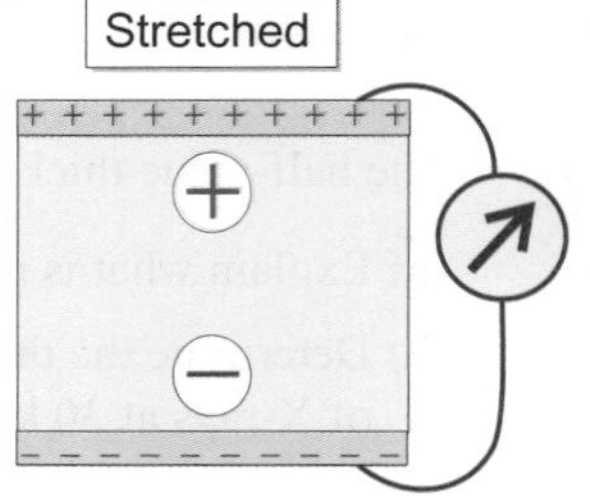

2) When you **apply a p.d.** across a piezoelectric crystal, the crystal **deforms**. If the p.d. is **alternating**, then the crystal **vibrates** at the **same frequency**.
3) A piezoelectric crystal can act as a **receiver** of **ultrasound**, converting **sound waves** into **alternating voltages**, and also as a **transmitter**, converting **alternating voltages** into **sound waves**.
4) Ultrasound devices use **lead zirconate titanate** (**PZT**) crystals. The **thickness** of the crystal is **half the wavelength** of the ultrasound that it produces. Ultrasound of this frequency will make the crystal **resonate** (like air in an open pipe — see p.12) and produce a large signal.
5) The PZT crystal is **heavily damped**, to produce **short pulses** and **increase** the **resolution** of the device.

Ultrasound Imaging

You need a Coupling Medium between the Transducer and the Body

1) **Soft tissue** has a very different **acoustic impedance** from **air**, so almost all the ultrasound **energy** is **reflected** from the surface of the body if there is air between the **transducer** and the **body**.
2) To avoid this, you need a **coupling medium** between the transducer and the body — this **displaces** the **air** and has an impedance much closer to that of body tissue. The use of **coupling media** is an example of **impedance matching**.
3) The coupling medium is usually an **oil** or **gel** that is smeared onto the skin.

The A-Scan is a Range Measuring System

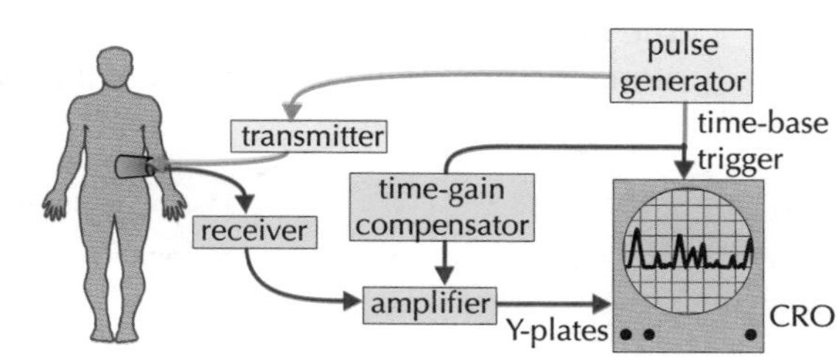

1) The **amplitude scan** (**A-Scan**) sends a short **pulse** of ultrasound into the body simultaneously with an **electron beam** sweeping across a cathode ray oscilloscope (**CRO**) screen.
2) The scanner receives **reflected** ultrasound pulses that appear as **vertical deflections** on the CRO screen.
3) **Weaker** pulses (that have travelled further in the body and **arrive later**) are **amplified** more to avoid the loss of valuable data — this process is called **time-gain compensation** (**TGC**).
4) The **horizontal positions** of the reflected pulses indicate the **time** the 'echo' took to return, and are used to work out **distances** between structures in the body (e.g. the **diameter** of a **baby's head** in the uterus).
5) A **stream** of pulses can produce a **steady image** on the screen (due to **persistence of vision** — see p.71), although modern CROs can store a digital image after just one exposure.

In a B-Scan, the Brightness Varies

1) In a **brightness scan** (**B-Scan**), the electron beam sweeps **down** the screen rather than across.
2) The amplitude of the reflected pulses is displayed as the **brightness** of the spot.
3) You can use a **linear array** of transducers to produce a **two-dimensional** image.

transducer probe
body
reflected signal
A
B
transmitted signal
on the CRO screen
A-Scan (rotated 90°)
B-Scan
reflection from A
reflection from B

Practice Questions

Q1 What are the main advantages and disadvantages of imaging using ultrasound?

Q2 How are ultrasound waves produced and received in an ultrasound transducer?

Q3 Define acoustic impedance.

Exam Questions

Q1 (a) What fraction of intensity is reflected when ultrasound waves pass from air to soft tissue? Use $Z_{air} = 0.430 \times 10^3$ kgm^{-2}s^{-1}, $Z_{tissue} = 1630 \times 10^3$ kgm^{-2}s^{-1}. [2 marks]

(b) Calculate the ratio between the intensity of the ultrasound that **enters** the body when a coupling gel is used ($Z_{gel} = 1500 \times 10^3$ kgm^{-2}s^{-1}) and when none is used. Give your answer to the nearest power of ten. [4 marks]

Q2 (a) The acoustic impedance of a certain soft tissue is 1.63×10^6 kgm^{-2}s^{-1} and its density is 1.09×10^3 kgm^{-3}. Show that ultrasound travels with a velocity of 1.50 kms^{-1} in this medium. [2 marks]

(b) The time base on a CRO was set to be 50 μscm^{-1}. Reflected pulses from either side of a fetal head are 2.4 cm apart on the screen. Calculate the diameter of the fetal head if the ultrasound travels at 1.5 kms^{-1}. [4 marks]

Ultrasound — Mancunian for 'très bien'

You can use ultrasound to make images in cases where X-rays would do too much damage — like to check up on the development of a baby in the womb. You have to know what you're looking for though, or it just looks like a blob.

Endoscopy

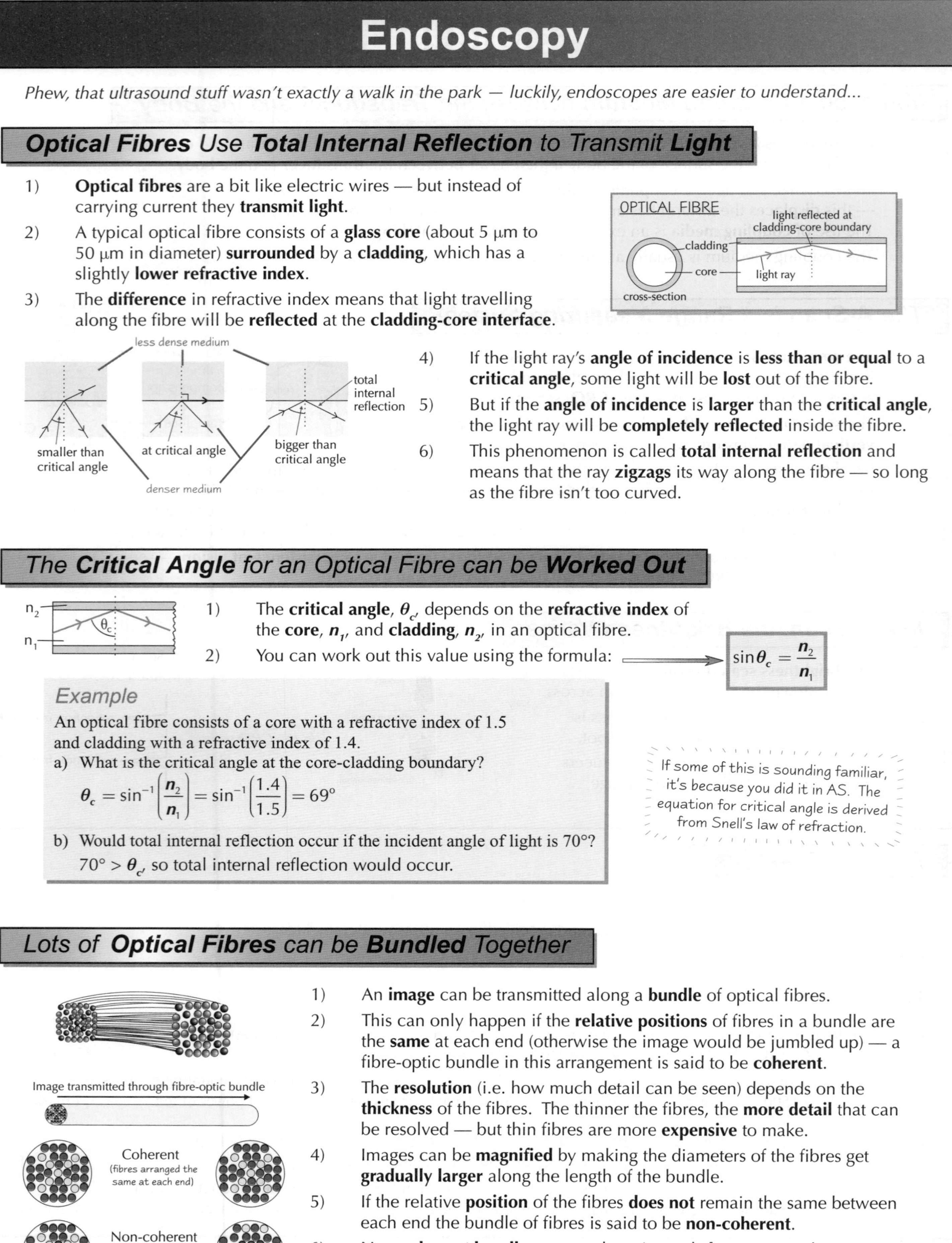

Phew, that ultrasound stuff wasn't exactly a walk in the park — luckily, endoscopes are easier to understand...

Optical Fibres Use *Total Internal Reflection* to Transmit *Light*

1) **Optical fibres** are a bit like electric wires — but instead of carrying current they **transmit light**.
2) A typical optical fibre consists of a **glass core** (about 5 μm to 50 μm in diameter) **surrounded** by a **cladding**, which has a slightly **lower refractive index**.
3) The **difference** in refractive index means that light travelling along the fibre will be **reflected** at the **cladding-core interface**.

4) If the light ray's **angle of incidence** is **less than or equal** to a **critical angle**, some light will be **lost** out of the fibre.
5) But if the **angle of incidence** is **larger** than the **critical angle**, the light ray will be **completely reflected** inside the fibre.
6) This phenomenon is called **total internal reflection** and means that the ray **zigzags** its way along the fibre — so long as the fibre isn't too curved.

The *Critical Angle* for an Optical Fibre can be *Worked Out*

1) The **critical angle**, θ_c, depends on the **refractive index** of the **core**, n_1, and **cladding**, n_2, in an optical fibre.
2) You can work out this value using the formula: $\sin\theta_c = \frac{n_2}{n_1}$

Example

An optical fibre consists of a core with a refractive index of 1.5 and cladding with a refractive index of 1.4.

a) What is the critical angle at the core-cladding boundary?

$$\theta_c = \sin^{-1}\left(\frac{n_2}{n_1}\right) = \sin^{-1}\left(\frac{1.4}{1.5}\right) = 69°$$

b) Would total internal reflection occur if the incident angle of light is 70°?

70° > θ_c, so total internal reflection would occur.

If some of this is sounding familiar, it's because you did it in AS. The equation for critical angle is derived from Snell's law of refraction.

Lots of *Optical Fibres* can be *Bundled* Together

1) An **image** can be transmitted along a **bundle** of optical fibres.
2) This can only happen if the **relative positions** of fibres in a bundle are the **same** at each end (otherwise the image would be jumbled up) — a fibre-optic bundle in this arrangement is said to be **coherent**.
3) The **resolution** (i.e. how much detail can be seen) depends on the **thickness** of the fibres. The thinner the fibres, the **more detail** that can be resolved — but thin fibres are more **expensive** to make.
4) Images can be **magnified** by making the diameters of the fibres get **gradually larger** along the length of the bundle.
5) If the relative **position** of the fibres **does not** remain the same between each end the bundle of fibres is said to be **non-coherent**.
6) **Non-coherent bundles** are much easier and **cheaper** to make. They **can't** transmit an **image** but they can be used to get **light** to hard-to-reach places — kind of like a flexible **torch**.

Endoscopy

Endoscopes Use Optical Fibres to Create an Image

1) An **endoscope** consists of a **long tube** containing **two bundles** of fibres — a **non-coherent** bundle to carry **light** to the area of interest and a **coherent** bundle to carry an **image** back to the eyepiece.
2) Endoscopes are widely used by surgeons to examine inside the body
3) An **objective lens** is placed at the **distal** end (**furthest from the eye**) of the **coherent** bundle to form an image, which is then transmitted by the fibres to the **proximal** end (**closest to the eye**) where it can be **viewed** through an **eyepiece**.
4) The **endoscope tube** can also contain a **water channel**, for cleaning the objective lens, a **tool aperture** to perform **keyhole surgery** and a CO_2 **channel** which allows CO_2 to be pumped into the area in front of the endoscope, making more room in the body.

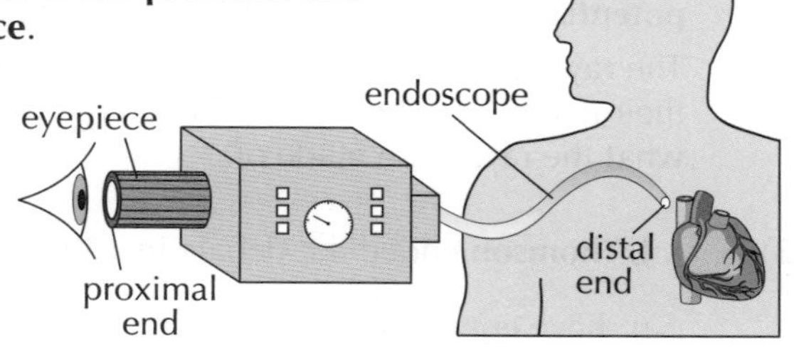

Endoscopes are Used in Keyhole Surgery

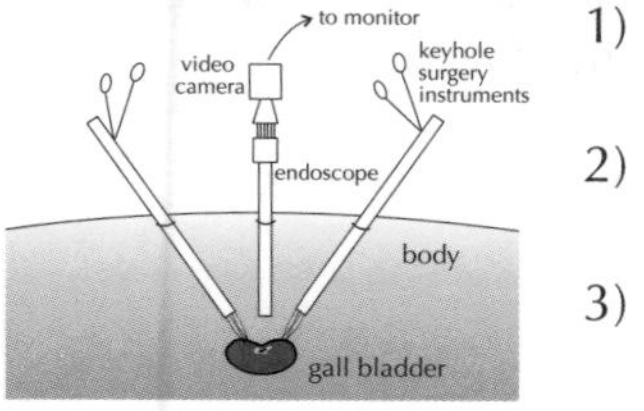

1) **Traditional** surgery needs a **large cut** to be made in the body so that there's **room** for the surgeons to get in and perform an **operation**.
2) This means that there's a **large risk of infection** to the exposed tissues and that permanent **damage** could be done to the patient's **body**.
3) New techniques in **minimally invasive surgery** (MIS or **keyhole surgery**) mean that only a **few small holes** need to be cut into the body.
4) An **endoscope** can be used in keyhole surgery to show the surgeon an **image** of the area of interest. **Surgical instruments** are passed through the endoscope tube, or through additional **small holes** in the body, so that the **operation** can be carried out.
5) **Common procedures** include the removal of the **gall bladder**, investigation of the **middle ear**, and removal of abnormal polyps in the **colon** so that they can be investigated for **cancer**.
6) **Recovery times** tend to be **quicker** for keyhole surgery, so the **patient** can usually **return home** on the **same day** — which makes it much **cheaper** for the hospital and **nicer** for the patient.

Practice Questions

Q1 What condition must be satisfied for total internal reflection to occur?

Q2 Explain the difference between a coherent and a non-coherent bundle of fibres.

Q3 What are the main features of an endoscope?

Q4 How have endoscopes revolutionised some surgical techniques?

Exam Questions

Q1 A beam of light is transmitted through an optical fibre.
The refractive index of the fibre's core is 1.35 and the refractive index of its cladding 1.30.
(a) Determine the critical angle for the core-cladding boundary. [1 mark]
(b) Explain why the angle of incidence of the beam of light should be kept at or above the critical angle. [2 marks]

Q2 Coherent fibre-optic bundles can be used to transmit images.
Describe the main features of the structure of a coherent fibre-optic bundle,
and explain why each feature is important for the bundle's function. [4 marks]

If you ask me, physics is a whole bundle of non-coherentness...

If this is all getting too much, and your brain is as fried as a pork chipolata, just remember the wise words of revision wisdom from the great Spike Milligan — Ying tong, ying tong, ying tong, ying tong, ying tong, iddly-I-po, iddly-I-po...

Charge/Mass Ratio of the Electron

***e/m** was known for quite a long time before anyone came up with a way to measure e or m separately.*

Cathode Ray** is an **Old-Fashioned** name for a Beam of **Electrons

1) The phrase '**cathode ray**' was first used in 1876, to describe the **glow** that appears on the wall of a discharge tube like the one in the diagram, when a **potential difference** is applied across the terminals.
2) The **rays** seemed to come from the **cathode** (hence their name) and there was a lot of argument about **what** the rays were made of.

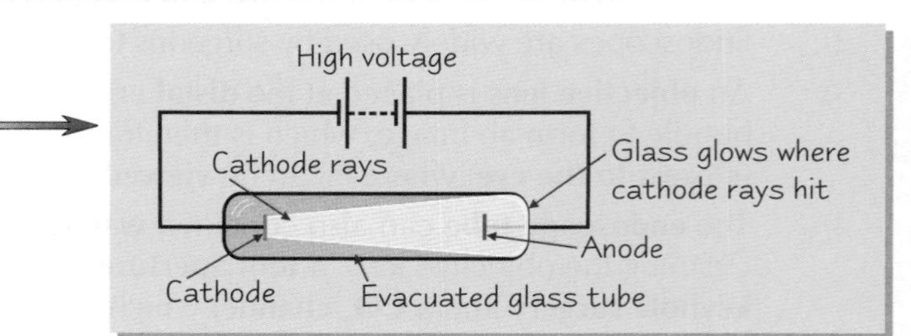

3) **J. J. Thomson** ended the debate in 1897, when he demonstrated (see opposite) that cathode rays:

   a) have **energy, momentum** and **mass**,
   b) have a **negative charge**,
   c) have the **same properties**, no matter **what gas** is in the tube and what the **cathode** is made of,
   d) have a **charge to mass ratio** much **bigger** than that of **hydrogen** ions. So they either have a **tiny mass**, or a much higher charge — Thomson assumed they had the same size charge as hydrogen ions.

   Thomson concluded that **all atoms** contain these 'cathode ray particles', that were later called **electrons**.

Electron Beams** are Produced by **Thermionic Emission

1) When you **heat** a **metal**, its **free electrons** gain a load of **thermal energy**.
2) Give them **enough energy** and they'll **break free** from the surface of the metal — this is called **thermionic emission**. (Try breaking the word down — think of it as '**therm**' [to do with heat] + '**ionic**' [to do with charge] + '**emission**' [giving off] — so it's 'giving off charged particles when you heat something'.)
3) Once they've been emitted, the electrons can be **accelerated** by an **electric field** in an **electron gun**:

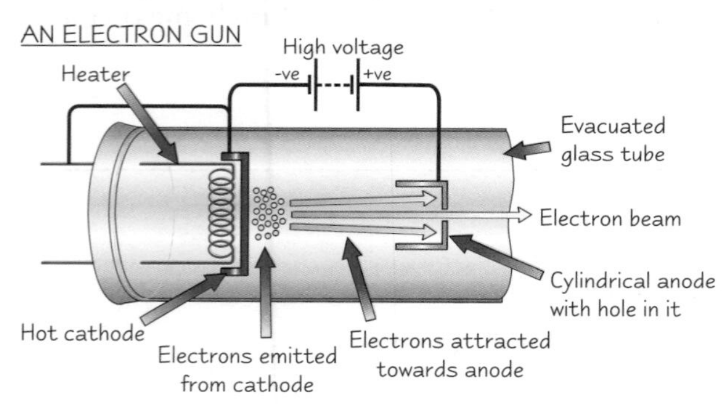

A **heating coil** heats the metal cathode. The electrons that are emitted are **accelerated** towards the **cylindrical anode** by the electric field set up by the high voltage.

Some electrons pass through a **little hole** in the **anode**, making a narrow electron beam. The electrons in the beam move at a **constant velocity** because there's **no field** beyond the anode — i.e., there's **no force**.

*The **Electronvolt** is Defined Using **Accelerated Charges***

1) The **kinetic energy** that a particle with charge ***Q* gains** when it's **accelerated** through a p.d. of ***V*** volts is ***QV*** joules. That just comes from the definition of the **volt** (JC^{-1}).
2) If you replace ***Q*** in the equation with the charge of a **single electron**, **e**, you get:

$$\frac{1}{2}mv^2 = eV$$

3) From this you can define a new **unit of energy** called the **electronvolt** (**eV**):

   1 electronvolt is the **kinetic energy carried** by an **electron** after it has been **accelerated** through a **potential difference** of **1 volt**.

4) So, the **energy in eV** of an electron accelerated by a potential difference is:

   energy gained by electron (eV) = accelerating voltage (V)

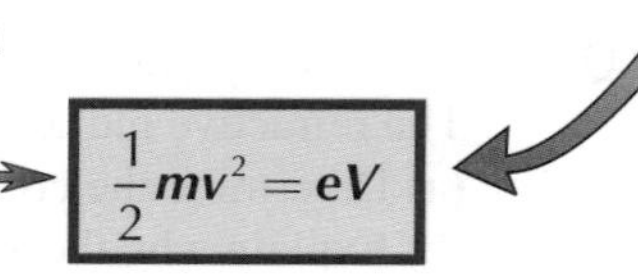

Conversion factor: $1 \text{ eV} = 1.6 \times 10^{-19} \text{ J}$

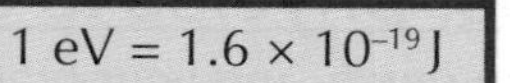

Charge/Mass Ratio of the Electron

Thomson Measured the **Specific Charge** of the **Electron**

1) The **specific charge** or **charge/mass ratio** of a charged particle is just its **charge** per unit **mass**.
2) There are a **few different ways** of measuring it, and you need to know about **one** of them. This isn't the method that Thomson used, but that's not important.

Check out Unit 4: Section 4 — Magnetic Fields, if you're having trouble with the experiment.

Measuring the Charge/Mass Ratio of an Electron

1) Electrons are charged particles, so they can be deflected by an **electric** or a **magnetic field**. This method uses a magnetic field in a piece of apparatus called a **fine beam tube**.
2) When the beam of electrons from the **electron gun** (see previous page) passes through the low-pressure gas, the hydrogen atoms along its path **absorb energy**. As the electrons in these **excited hydrogen atoms** fall back to the ground state, they **emit light**. The electron beam is seen as a **glowing trace** through the gas.
3) Two circular **magnetic field coils** either side generate a **uniform magnetic field** inside the tube.
4) The electron beam is initially fired at **right angles** to the **magnetic field**, so the beam curves round in a **circle**.

magnetic field coils
electron gun
electron beam
glass bulb containing hydrogen at low pressure

5) This means that the **magnetic force** on the electron (see p. 27) is acting as a **centripetal force** (see p. 7). So the radius of the circle is given by:

$$\frac{mv^2}{r} = Bev$$

where m is the mass of an electron, e is the charge on an electron, B is the magnetic field strength, v is the velocity of the electron and r is the radius of the circle.

6) From the previous page, you've got an equation that you can rearrange to give v in terms of the **accelerating potential** of the electron gun. If you substitute that expression for v into the equation above (and tidy it all up a bit) you get:

$$\frac{e}{m} = \frac{2V}{B^2r^2}$$

where m is the mass of an electron, e is the charge on an electron, B is the magnetic field strength, V is the accelerating potential and r is the radius of the circle.

You can **measure** all the quantities on the **right-hand side** of the equation using the **fine beam tube**, leaving you with the **specific charge, e/m**. It turns out that e/m (1.76×10^{11} Ckg^{-1}) is about **1800 times greater** than the **specific charge of a hydrogen ion** or **proton** (9.58×10^{7} Ckg^{-1}). And the **mass** of a **proton** is about **1800 times greater** than the **mass** of an **electron** — **Thomson was right**, electrons and protons do have the **same size charge**.

Practice Questions

Q1 What is meant by thermionic emission?

Q2 Sketch a labelled diagram of an electron gun that could be used to accelerate electrons.

Q3 What was Thomson's main conclusion following his measurement of e/m for electrons?

Exam Questions

Q1 An electron of mass 9.1×10^{-31} kg and charge -1.6×10^{-19} C is accelerated through a potential difference of 1 kV.

(a) State its energy in eV. [1 mark]

(b) Calculate its energy in joules. [1 mark]

(c) Calculate its speed in ms^{-1} and express this as a percentage of the speed of light (3.0×10^{8} ms^{-1}). [3 marks]

Q2 Explain the main features of an experiment to determine the specific charge of the electron.
The quality of your written answer will be assessed in this question. [5 marks]

New Olympic event — the electronvault...

Electronvolts are really handy units — they crop up all over the rest of this book, particularly in nuclear and particle physics. It stops you having to mess around with a load of nasty powers of ten. Cathode ray tubes (CRTs) are pretty handy too — there might be one in your telly... unless it's one of those new-fangled flat-screen plasma thingummy-do-dahs...

Millikan's Oil-Drop Experiment

Thomson had already found the charge/mass ratio of the electron in 1897 — now it was down to Robert Millikan, experimenter extraordinaire, to find the absolute charge...

Millikan's Experiment used Stoke's Law

1) Before you start thinking about Millikan's experiment, you need a bit of **extra theory**.
2) When you drop an object into a fluid, like air, it experiences a **viscous drag** force. This force acts in the **opposite direction** to the velocity of the object, and is due to the **viscosity** of the fluid.
3) You can calculate this viscous force on a spherical object using **Stoke's law**:

$$F = 6\pi\eta r v$$

where η is the viscosity of the fluid, r is the radius of the object and v is the velocity of the object.

Millikan's Experiment — the Basic Set-Up

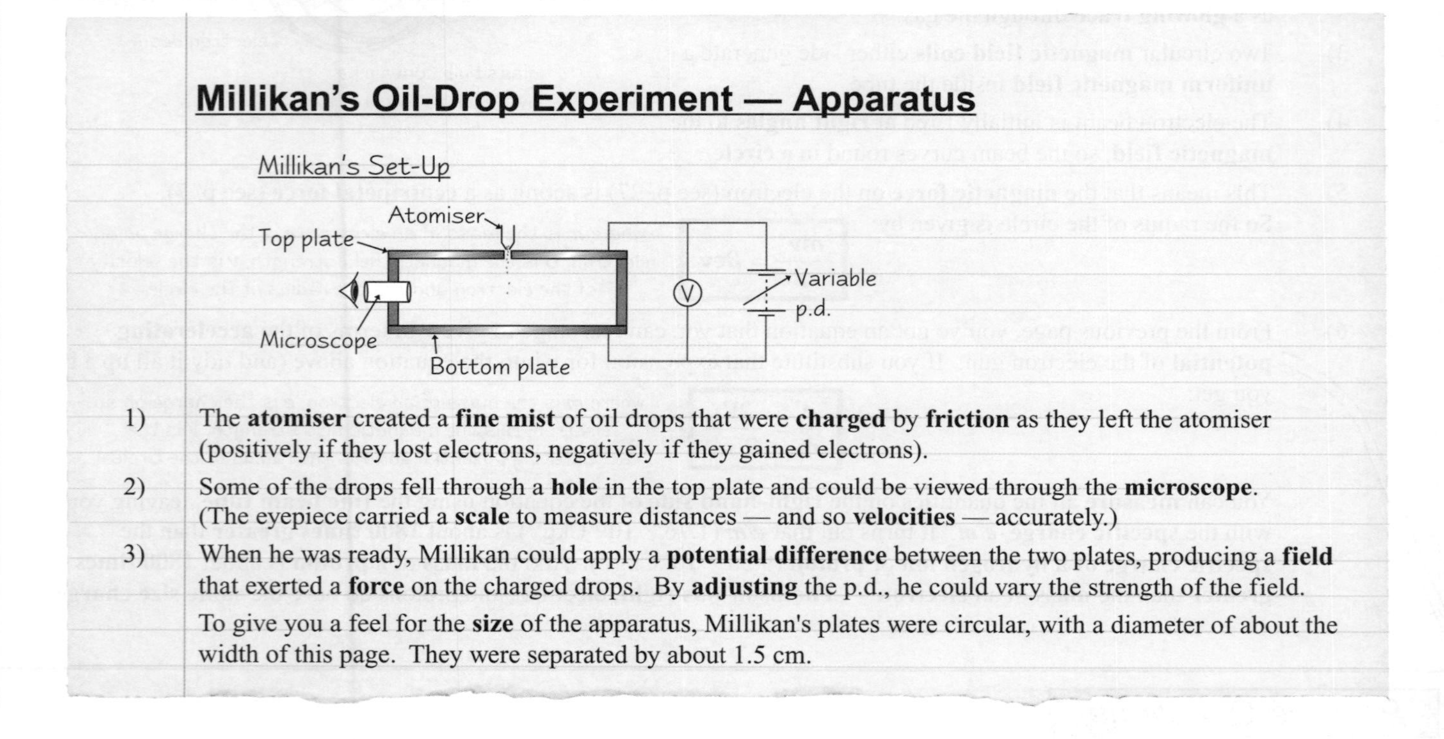

Millikan's Oil-Drop Experiment — Apparatus

1) The **atomiser** created a **fine mist** of oil drops that were **charged** by **friction** as they left the atomiser (positively if they lost electrons, negatively if they gained electrons).
2) Some of the drops fell through a **hole** in the top plate and could be viewed through the **microscope**. (The eyepiece carried a **scale** to measure distances — and so **velocities** — accurately.)
3) When he was ready, Millikan could apply a **potential difference** between the two plates, producing a **field** that exerted a **force** on the charged drops. By **adjusting** the p.d., he could vary the strength of the field.

To give you a feel for the **size** of the apparatus, Millikan's plates were circular, with a diameter of about the width of this page. They were separated by about 1.5 cm.

Before the Field is Switched on, there's only Gravity and the Viscous Force

1) With the electric field turned off, the forces acting on each oil drop are:
   a) the **weight** of the drop — acting downwards
   b) the **viscous force** from the air — acting upwards

Millikan had to take account of things like upthrust as well, but you don't have to worry about that — keep it simple.

2) The drop will reach **terminal velocity** (i.e. it will stop accelerating) when these two forces are equal. So, from Stoke's law (see above):

$$mg = 6\pi\eta r v$$

3) Since the **mass** of the drop is the **volume** of the drop multiplied by the **density**, ρ, of the oil, this can be rewritten as:

$$\frac{4}{3}\pi r^3 \rho g = 6\pi\eta r v \quad \Rightarrow \quad r^2 = \frac{9\eta v}{2\rho g}$$

Millikan measured η and ρ in separate experiments, so he could now calculate r — ready to be used when he switched on the electric field...

Millikan's Oil-Drop Experiment

Then he **Turned On** the **Electric Field...**

1) The field introduced a **third major factor** — an **electric force** on the drop.
2) Millikan adjusted the applied p.d. until the drop was **stationary**. Since the **viscous force** is proportional to the **velocity** of the object, once the drop stopped moving, the viscous force **disappeared**.
3) Now the only two forces acting on the oil drop were:

    a) the **weight** of the drop — acting downwards
    b) the force due to the **uniform electric field** — acting upwards

4) The **electric force** is given by:

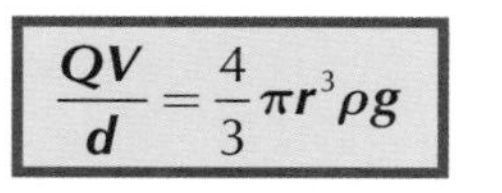

$$F = \frac{QV}{d}$$

where Q is the charge on the oil drop, V is the p.d. between the plates and d is the distance between the plates.

See p. 18-20

5) Since the drop is **stationary**, this electric force must be equal to the weight, so:

$$\frac{QV}{d} = \frac{4}{3}\pi r^3 \rho g$$

The first part of the experiment gave a value for r, so the **only unknown** in this equation is Q.

6) So Millikan could find the **charge on the drop**, and repeated the experiment for hundreds of drops. The charge on any drop was always a **whole number multiple** of **-1.6×10^{-19} C**.

These Results Suggested that **Charge** was **Quantised**

1) This result was **really significant**. Millikan concluded that charge can **never exist** in **smaller** quantities than 1.6×10^{-19} C. He assumed that this was the **charge** carried by an **electron**.
2) Later experiments confirmed that **both** these things are true.

> Charge is "**quantised**". It exists in "packets" of size **1.6×10^{-19} C** — the **fundamental unit of charge**. This is the size of the charge carried by **one electron**.

Practice Questions

Q1 Write down the equation for Stoke's law, defining any variables.

Q2 List the forces that act on the oil drop in Millikan's experiment:
(a) with the drop drifting downwards at terminal velocity but with no applied electrical field,
(b) when the drop is stationary, with an electrical field applied.

Q3 Briefly explain the significance of Millikan's oil-drop experiment in the context of quantum physics.

Exam Question

Q1 An oil drop of mass 1.63×10^{-14} kg is held stationary in the space between two charged plates 3.00 cm apart. The potential difference between the plates is 5000 V. The density of the oil used is 880 kgm^{-3}.

(a) Describe the relative magnitude and direction of the forces acting on the oil drop. [2 marks]

(b) Calculate the charge on the oil drop using $g = 9.81$ Nkg^{-1}.
Give your answer in terms of e, the charge on an electron. [3 marks]

The electric field is switched off and the oil drop falls towards the bottom plate.

(c) Explain why the oil drop reaches terminal velocity as it falls. [3 marks]

(d) Calculate the terminal velocity of the oil drop using $\eta = 1.84 \times 10^{-5}$ kgm^{-1}s^{-1}. [3 marks]

So next time you've got a yen for 1.59×10^{-19} coulombs — tough...

This was a huge leap. Along with the photoelectric effect (see p. 94) this experiment marked the beginning of quantum physics. The world wasn't ruled by smooth curves any more — charge now jumped from one allowed step to the next...

Light — Newton vs Huygens

Newton was quite a bright chap really, but even he could make mistakes — and this was his biggest one. The trouble with being Isaac Newton is that everyone just assumes you're right...

Newton had his Corpuscular Theory

1) In 1672, Newton published his **theory of colour**. In it he suggested that **light** was made up of **tiny particles** that he called '**corpuscles**'.
2) One of his major arguments was that light was known to travel in **straight lines**, yet waves were known to **bend** in the shadow of an **obstacle** (diffraction). Experiments weren't **accurate enough** then to detect the diffraction of light. Light was known to **reflect** and **refract**, but that was it.
3) His theory was based on the principles of his **laws of motion** — that all particles, including his 'corpuscles', will 'naturally' travel in **straight lines**.
4) Newton believed that **reflection** was due to a force that **pushed** the particles away from the surface — just like a ball bouncing back off a wall.
5) **Refraction** worked if the corpuscles travelled **faster** in a **denser** medium.

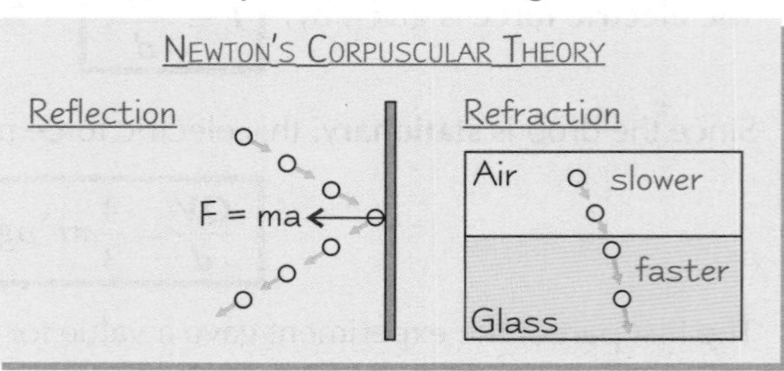

Huygens thought Light was a Wave

1) The idea that light might be a **wave** had existed for some time before it was formalised by Huygens in 1678.
2) At the time, nobody took much notice of him because his theory was **different** from Newton's.
3) Huygens developed a **general model** of the propagation of **waves** in what is now known as **Huygens' principle**:

> **HUYGENS' PRINCIPLE:** Every point on a wavefront may be considered to be a **point source** of **secondary wavelets** that spread out in the forward direction at the speed of the wave. The **new wave front** is the surface that is **tangential** to all of these **secondary wavelets**.

The diagram below shows how this works:

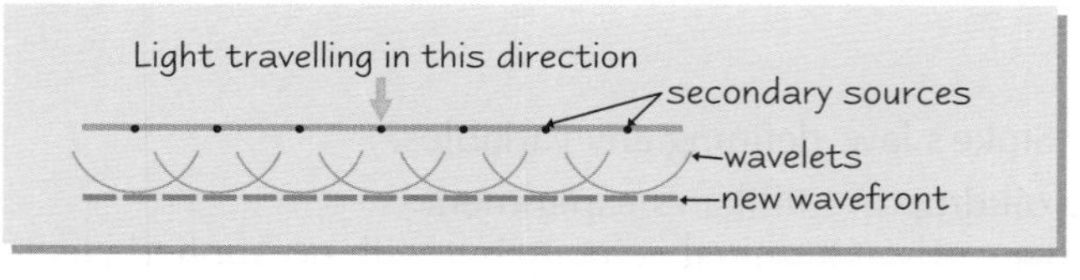

4) By applying his theory to **light**, he found that he could explain **reflection** and **refraction** easily.
   Huygens predicted that light should **slow down** when it entered a **denser medium**, rather than speed up.

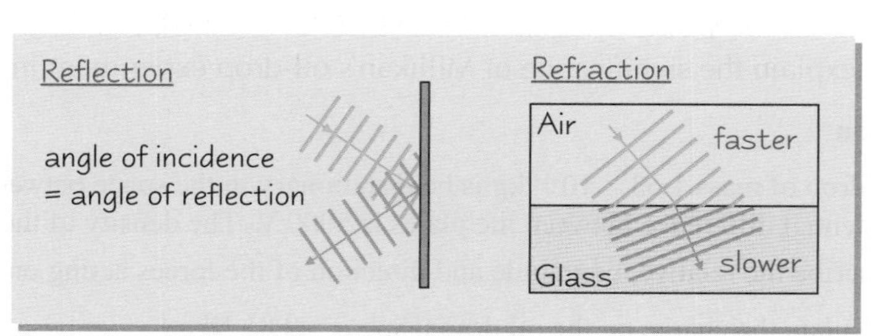

5) Huygens also predicted that light should **diffract** around tiny objects and that two coherent light sources should **interfere** with each other.

> Up until the end of the 18th century, most scientists sided with **Newton**. He'd been right about so many things before, so it was generally assumed that he **must be right** about light being corpuscular. The debate raged for **over 100 years** until **Thomas Young** carried out experiments on the **interference** of light in Cambridge around **1800**...

Light — Newton vs Huygens

Young Proved Huygens Right with his Double-Slit Experiment

1) **Diffraction** and **interference** are both uniquely **wave** properties. If it could be shown that **light** showed **interference** patterns, that would help decide once and for all between corpuscular theory and wave theory.
2) The problem with this was getting two **coherent** light sources, as **light** is emitted in **random bursts**.
3) Young solved this problem by using only **one point source of light** (a light source behind a narrow slit). In front of this was a **screen** with **two narrow slits** in it. Light spreading out by **diffraction** from the slits was equivalent to **two coherent point sources**.
4) In the area on the screen where light from the two slits **overlapped**, bright and dark '**fringes**' were formed. This was **proof** that light could both **diffract** (through the narrow slits) and **interfere** (to form the interference pattern on the screen) — **Huygens** was right all along.

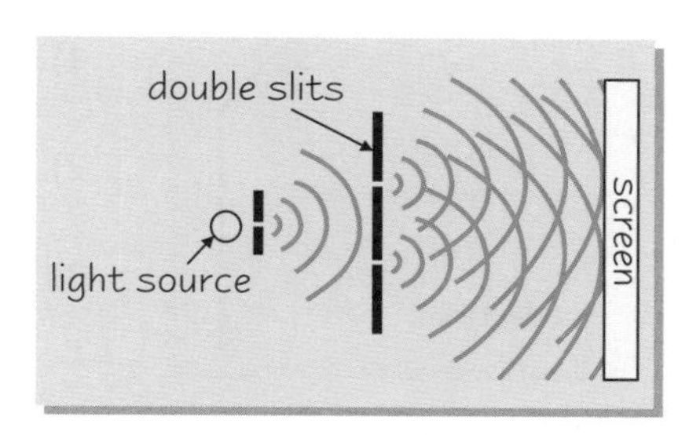

Observations and Theories Developed Rapidly during the 19th Century

1) In 1808, Etienne-Louis Malus discovered that light was **polarised** by reflection. Physicists at the time thought that light spread like sound, as a longitudinal wave, so they struggled to explain polarisation.
2) In 1817, Young suggested that light was a **transverse wave** consisting of **vibrating electric** and **magnetic fields** at **right angles** to each other and the **direction of travel**. This explained why light could be **polarised**.
3) In the second half of the 19th century, James Clerk Maxwell showed theoretically that **all electromagnetic waves** should travel at the same speed in a vacuum, **c**.

> **James Clerk Maxwell** calculated the **speed of light** in a vacuum using:
>
> $$c = \frac{1}{\sqrt{\mu_0 \varepsilon_0}}$$
>
> where **c** is the speed of the wave in ms^{-1}, μ_0 ("mu-nought") is the permeability of free space (a constant — $4\pi \times 10^{-7}$ Hm^{-1}) and ε_0 ("epsilon-nought") is the permittivity of free space (another constant — 8.85×10^{-12} Fm^{-1}).
>
> μ_0 relates to the **magnetic flux density** due to a current-carrying wire in free space, while ε_0 relates to the **electric field strength** due to a charged object in free space.

4) By that time, the **velocity of light** could be measured quite accurately, and it was found to be very close to Maxwell's value of **c**. This suggested that **light** is an **electromagnetic wave**. We now know that all electromagnetic waves, including light, travel in a **vacuum** at a **speed** of **2.998×10^8 ms^{-1}**.
5) In 1887, **Heinrich Hertz** produced and detected **radio waves** using electric sparks. He showed by **experiment** that they could be reflected, refracted, diffracted and polarised, and show interference. This helped confirm that radio waves, like light, are electromagnetic waves.
6) This was the accepted theory up until the very end of the 19th century, when the **photoelectric effect** was discovered. Then the particle theory had to be resurrected, and it was all up in the air again...

Practice Questions

Q1 What was the main argument that Newton used to support his corpuscular theory of light?

Q2 What part does diffraction play in a Young's double-slit experiment?

Q3 Sketch a diagram showing an experiment to demonstrate Young's fringes for white light in a laboratory.

Exam Questions

Q1 Describe Newton's corpuscular theory of light. [2 marks]

Q2 Give a brief history of the understanding of the nature of light in the 18th and 19th centuries. You should mention the various theories that have been proposed and the evidence that has been used to support them.
The quality of your written answer will be assessed in this question. [6 marks]

In the blue corner — the reigning champion... IsaaaAAAAC NEWton...

So, light's a wave, right? We've got that sorted — the double-slit experiment laid the whole argument to rest... or did it?

The Photoelectric Effect

You did the photoelectric effect at AS, so it should be familiar. You need it again though, so here it is...

Shining Light on a *Metal* can *Release Electrons*

If you shine **light** of a **high enough frequency** onto the **surface of a metal**, it will **emit electrons**. For **most** metals, this **frequency** falls in the **UV** range.

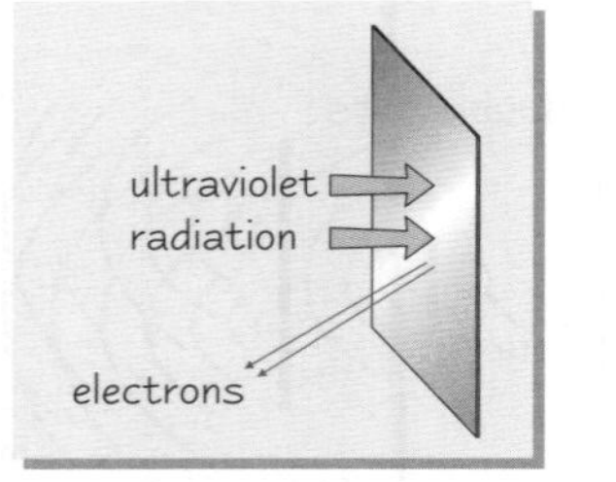

1) **Free electrons** on the **surface** of the metal **absorb energy** from the light, making them **vibrate** or move faster.
2) If an electron **absorbs enough** energy, the **bonds** holding it to the metal can be **broken** and the electron **released**.
3) This is called the **photoelectric effect** and the electrons emitted are called **photoelectrons**.

You don't need to know the details of any experiments on this — you just need to learn the three main conclusions:

Conclusion 1	For a given metal, **no photoelectrons are emitted** if the radiation has a frequency **below** a certain value — called the **threshold frequency**.
Conclusion 2	The photoelectrons are emitted with a variety of kinetic energies ranging from zero to some maximum value. This value of **maximum kinetic energy** increases with the **frequency** of the radiation, and is **unaffected** by the **intensity** of the radiation.
Conclusion 3	The **number** of photoelectrons emitted per second is **directly proportional** to the **intensity** of the radiation.

These are the two that had people puzzled. They can't be explained using wave theory.

The *Photoelectric Effect Couldn't* be Explained by *Wave Theory*

According to wave theory:

1) For a particular frequency of light, the **energy** carried is **proportional** to the **intensity** of the beam.
2) The energy carried by the light would be **spread evenly** over the wavefront.
3) **Each** free electron on the surface of the metal would gain a **bit of energy** from each incoming wave.
4) Gradually, each electron would gain **enough energy** to be able to leave the metal.

SO... If the light had a **lower frequency** (i.e. was carrying less energy) it would take **longer** for the electrons to gain enough energy — but it would happen eventually. There is **no explanation** for the **threshold frequency**.

The **higher the intensity** of the wave, the **more energy** it should transfer to each electron — so the kinetic energy should increase with **intensity**. There's **no explanation** for the **kinetic energy** depending only on the **frequency**.

Einstein came up with the *Photon Model of Light*

1) When Max Planck was investigating **black body radiation** (don't worry, you don't need to know about that just yet), he suggested that **EM waves** can **only** be **released** in **discrete packets**, or **quanta**.
2) The **energy carried** by one of these **wave-packets** had to be:

$$E = hf = \frac{hc}{\lambda}$$

where h = Planck's constant = 6.63×10^{-34} Js
and c = speed of light in a vacuum = 3.00×10^{8} ms^{-1}

3) **Einstein** went **further** by suggesting that **EM waves** (and the energy they carry) can only **exist** in discrete packets. He called these wave-packets **photons**.
4) He saw these photons of light as having a **one-on-one**, **particle-like** interaction with **an electron** in a **metal surface**. It would **transfer all** its **energy** to that **one**, **specific electron**.

The Photoelectric Effect

The **Photon Model** Explained the **Photoelectric Effect** Nicely

According to the photon model, when light hits its surface, the metal is **bombarded** by photons. If one of these photons **collides** with a free electron, the electron will gain energy equal to ***hf***.

Before an electron can **leave** the surface of the metal, it needs enough energy to **break the bonds holding it there**. This energy is called the **work function** (symbol ϕ) and its **value** depends on the **metal**.

It Explains the **Threshold Frequency...**

1) If the energy **gained** from the photon is **greater** than the **work function**, the electron can be **emitted**.
2) If it **isn't**, the electron will just **shake about a bit** or move faster, then release the energy as another photon or in collisions. The metal will heat up, but **no electrons** will be emitted.
   Since for **electrons** to be released, $hf \geq \phi$, the **threshold frequency** must be:

$$f = \frac{\phi}{h}$$

... and the **Maximum Kinetic Energy**

1) The **energy transferred** to an electron is ***hf***.
2) The **kinetic energy** it will be carrying when it **leaves** the metal is ***hf* minus** any energy it's **lost** on the way out (there are loads of ways it can do that, which explains the **range** of energies).
3) The **minimum** amount of energy it can lose is the **work function**, so the **maximum kinetic energy** is given by the equation:
   The **kinetic energy** of the electrons is **independent of intensity**, as they can **only absorb one photon** at a time.

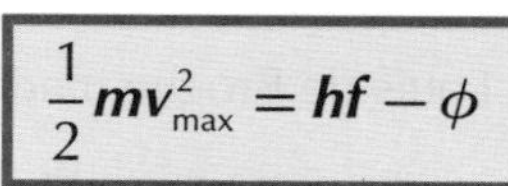

$$\frac{1}{2}mv_{max}^2 = hf - \phi$$

4) The maximum kinetic energy can be **measured** using the idea of **stopping potential**. The emitted electrons are made to lose their **energy** by **doing work against** an applied **potential difference**. The work done by the p.d. in **stopping** the **fastest electrons** is equal to the energy they were carrying:

$$\frac{1}{2}mv_{max}^2 = eV_s$$

where **e** = charge on the electron = 1.6×10^{-19} C, V_s = stopping potential in V, and the maximum kinetic energy is measured in J. The kinetic energy in eV is just equal to the stopping potential — see p. 88.

5) The stopping potential for a frequency of light can be measured using a circuit containing a **vacuum photocell** (two metal plates in a vacuum). The light is directed on to one plate, causing **photoelectrons** to be released.
6) If the photoelectrons reach the other plate, a **current** flows in the circuit. The p.d. across the plates is **increased** from 0 using a **potential divider.** When the current **stops**, the **stopping potential** has been reached.

Practice Questions

Q1 What three main conclusions were drawn from detailed experimentation on the photoelectric effect?

Q2 What is meant by the work function energy of a metal?

Q3 How is the energy of a photon related to its frequency?

Exam Questions *Use Planck's constant = 6.63×10^{-34} Js, speed of light = 3.00×10^{8} ms^{-1}, electron charge = 1.6×10^{-19} C.*

Q1 An isolated zinc plate with neutral charge is exposed to high-frequency ultraviolet light. State and explain the effect of the ultraviolet light on the charge of the plate. [2 marks]

Q2 Light of wavelength 0.50 μm hits a metal surface and causes electrons to be released. Their maximum kinetic energy is 2.0×10^{-19} J.

(a) Calculate the work function of the metal. [3 marks]

(b) Explain why a beam of light with a wavelength of 1.5 μm would not cause electrons to be emitted. [3 marks]

Q3 Potassium has a work function of 2.2 eV. A potassium anode is illuminated with light of wavelength 350 nm. What potential must be applied to stop the electrons leaving the anode? [4 marks]

And that's all there is to it — *sob*...

Learn why the light theory can't explain the photoelectric effect, and how photon theory does — then tell someone else.

Wave-Particle Duality

Is it a wave? Is it a particle?

Interference and Diffraction show Light as a Wave

1) Light produces **interference** and **diffraction** patterns — **alternating bands** of **dark** and **light**.
2) These can **only** be explained using **waves interfering constructively** (when two waves overlap in phase) or **interfering destructively** (when two waves are out of phase).

The Photoelectric Effect Shows Light Behaving as a Particle

1) **Einstein** explained the results of **photoelectricity experiments** (see p. 94) by thinking of the **beam of light** as a series of **particle-like "photons"**.
2) If a **photon** of light is a **discrete** bundle of energy, then it can **interact** with an **electron** in a **one-to-one way**.
3) **All** the **energy** in the **photon** is **given** to one **electron**.

Neither the **wave theory** nor the **particle theory** describe what light actually **is**. They're just two different **models** that help to explain the way light behaves.

I'm not impressed — this is just speculation. What do you think Dad?

De Broglie came up with the Wave-Particle Duality Theory

1) Louis de Broglie made a **bold suggestion** in his **PhD thesis**:

> If "**wave-like**" **light** showed **particle properties** (photons), "**particles**" like **electrons** should be expected to show **wave-like properties**.

2) The **de Broglie equation** relates a **wave property** (**wavelength**, λ) to a **moving particle property** (**momentum**, mv). h = Planck's constant = 6.63×10^{-34} Js.

$$\lambda = \frac{h}{mv}$$

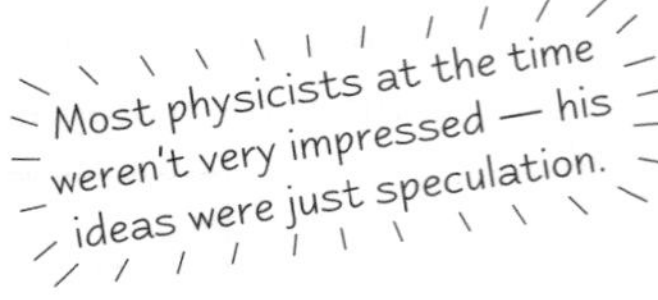

3) The **de Broglie wave** of a particle can be interpreted as a **"probability wave"**.
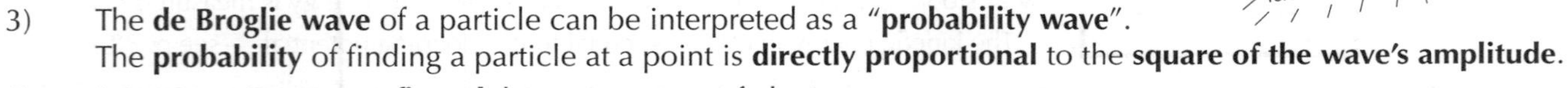
The **probability** of finding a particle at a point is **directly proportional** to the **square of the wave's amplitude**.
4) Later experiments **confirmed** the wave nature of electrons.

Electron Diffraction shows the Wave Nature of Electrons

1) De Broglie's suggestions prompted a lot of experiments to try to show that **electrons** can have **wave-like** properties. In **1927**, Davisson and Germer succeeded in **diffracting electrons**.
2) They saw **diffraction patterns** when **accelerated electrons** in a vacuum tube **interacted** with the **spaces** in a graphite **crystal**.
3) According to wave theory, the **spread** of the **lines** in the diffraction pattern **increases** if the **wavelength** of the wave **increases**.
4) In electron diffraction experiments, a **small accelerating voltage**, i.e. **slow** electrons, gives **widely spaced** rings.
5) **Increase** the **electron speed** and the diffraction pattern circles **squash together** towards the **middle**. This fits in with the **de Broglie** equation above — if the **velocity** is **higher**, the **wavelength** is **shorter** and the **spread** of the lines is **smaller**.

Electron diffraction patterns look like this.

For astrophysics students, this circle is called the Airy disc (see p. 53).

> In general, λ for **electrons** accelerated in a **vacuum tube** is about the **same size** as λ for **electromagnetic waves** in the **X-ray** part of the spectrum.

6) The de Broglie wavelength of an electron (λ) is related to the **accelerating voltage** (V) by:

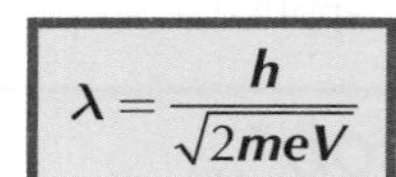

$$\lambda = \frac{h}{\sqrt{2meV}}$$

where e is the charge on the electron and m is its mass

Wave-Particle Duality

Particles Don't Show Wave-Like Properties All the Time

You **only** get **diffraction** if a particle interacts with an object of about the **same size** as its **de Broglie wavelength**.
A **tennis ball**, for example, with **mass 0.058 kg** and **speed 100 ms^{-1}** has a **de Broglie wavelength** of **10^{-34} m**.
That's **10^{19} times smaller** than the **nucleus** of an **atom**! There's nothing that small for it to interact with.

Example — An electron of mass 9×10^{-31} kg is fired from an electron gun at 7×10^{6} ms^{-1}. What size object will the electron need to interact with in order to diffract?

Momentum of electron = $mv = 6.3 \times 10^{-24}$ kg ms^{-1}

$\lambda = h/mv = 6.63 \times 10^{-34} / 6.3 \times 10^{-24} = 1 \times 10^{-10}$ m

Only crystals with atom layer spacing around this size are likely to cause the diffraction of this electron.

You can also calculate the anode voltage needed to produce this wavelength using the formula at the bottom of the previous page: $\lambda = \frac{h}{\sqrt{2meV}} \Rightarrow V = \frac{h^2}{2me\lambda^2} = \mathbf{153\ V}$

A **shorter wavelength** gives **less diffraction**. This is important in **microscopes** where diffraction **blurs out details**.
The **tiny** wavelength of electrons means an **electron microscope** can resolve **finer detail** than a **light** microscope.

Electron Microscopes use Electrons Instead of Light

In electron microscopes:

1) A **stream of electrons** is accelerated towards the sample using a **positive electric potential** — an **electron gun**.
2) To **resolve detail** around the size of an **atom**, the **electron wavelength** needs to be similar to the **diameter** of an **atom** (**0.1 nm**) — which (using the equation in the example above) means an **anode voltage** of **at least 150 V**.
3) The **stream of electrons** from the electron gun is confined into a thin **beam** using a **magnetic field**.
4) The beam is **focused** onto the sample and any interactions are transformed into an **image**.
The sort of image you get depends on the **type of microscope** you're using:

A **transmission electron microscope** (**TEM**) works a bit like a **slide** projector, but uses electrons instead of light. A **very thin** specimen is used and the parts of the beam that pass through the specimen are projected onto a **screen** to form an image.

A **scanning tunnelling microscope** (**STM**) is a different kind of microscope that uses principles of **quantum mechanics**. A very fine **probe** is moved over the surface of the sample and a **voltage** is applied between the probe and the surface. Electrons "**tunnel**" from the probe to the surface, resulting in a weak **electrical current**. The smaller the **distance** between the probe and the surface, the **greater the current**. By scanning the probe over the surface and measuring the current, you produce an **image** of the **surface** of the sample.

Practice Questions

Q1 What name is normally given to "particles" of light?
Q2 What observation showed that electrons could behave as waves?
Q3 What is the advantage of an electron microscope over a light microscope?

Exam Questions *Use $h = 6.63 \times 10^{-34}$ Js, $e = 1.6 \times 10^{-19}$ C, $m_e = 9.1 \times 10^{-31}$ kg.*

Q1 An electron is accelerated through a p.d. of 500 V.
(a) Calculate:
i) the velocity of the electron, ii) its de Broglie wavelength. [4 marks]
(b) In which region of the electromagnetic spectrum does this fall? [1 mark]

Q2 (a) Describe how a transmission electron microscope (TEM) uses a beam of electrons to produce an image. [3 marks]
(b) Show that an anode voltage of at least 150 V is needed for a TEM to resolve detail around the size of an atom (0.1 nm). [3 marks]

Wave-Particle duelity — pistols at dawn...

*You're getting into the weird bits of quantum physics now — it says that light isn't a wave, and it isn't a particle, it's **both**... at the **same time**. And if you think that's confusing, just wait till you get onto relativity — not that I want to put you off.*

The Speed of Light and Relativity

First — a bit of a history lesson.

Michelson and *Morley* tried to find the *Absolute Speed* of the Earth

1) During the 19th century, most physicists believed in the idea of **absolute motion**. They thought everything, including light, moved relative to a **fixed background** — something called the **ether**.
2) **Michelson** and **Morley** tried to measure the **absolute speed** of the **Earth** through the ether using a piece of apparatus called an **interferometer**.
3) They expected the motion of the Earth to affect the **speed of light** they measured in **certain directions**. According to Newton, the speed of light measured in a **lab** moving parallel to the light would be ($c + v$) or ($c - v$), where v is the speed of the lab. By measuring the speed of light **parallel** and **perpendicular** to the motion of the Earth, Michelson and Morley hoped to find v, the absolute speed of the Earth.

They used an *Interferometer* to Measure the Speed of the Earth

The interferometer was basically **two mirrors** and a **partial reflector** (a beam-splitter). When you shine light at a partial reflector, some of the light is **transmitted** and the rest is **reflected**, making **two separate beams**.

The mirrors were at **right angles** to each other, and an **equal distance**, L, from the beam-splitter.

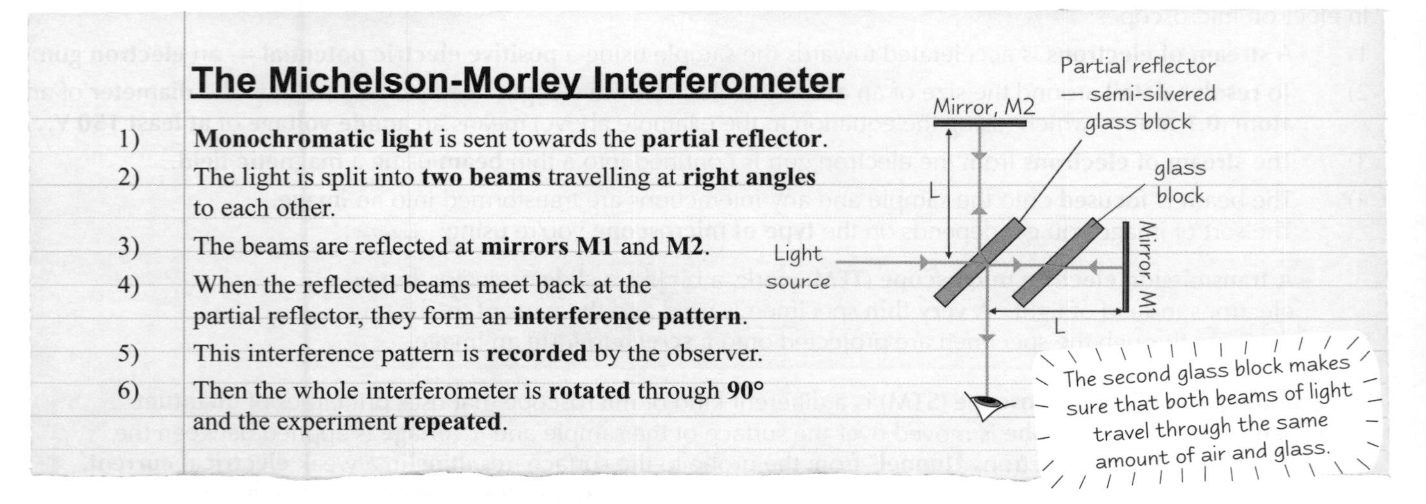

The Michelson-Morley Interferometer

1) **Monochromatic light** is sent towards the **partial reflector**.
2) The light is split into **two beams** travelling at **right angles** to each other.
3) The beams are reflected at **mirrors M1** and **M2**.
4) When the reflected beams meet back at the partial reflector, they form an **interference pattern**.
5) This interference pattern is **recorded** by the observer.
6) Then the whole interferometer is **rotated** through **90°** and the experiment **repeated**.

EXPECTED OUTCOME

According to Newton's laws, light moving **parallel** to the motion of the Earth should take **longer** to travel to the mirror and back than light travelling at **right angles** to the Earth's motion. So **rotating** the apparatus should have changed the **travel time** for the two beams.

This would cause a **tiny shift** in the **interference pattern**.

They *Didn't* get the *Result* they were *Expecting*

They **repeated** the experiment **over** and **over** again — at different **times of day** and at different points in the **year**. Taking into account any **experimental errors**, there was **absolutely no shift** in the interference pattern.

The time taken by each beam to travel to each mirror was **unaffected** by rotating the apparatus.

So, Newton's laws **didn't work** in this situation.

Most scientists were really puzzled by this "null result". Eventually, the following **conclusions** were drawn:

a) It's **impossible** to detect **absolute motion** — the ether doesn't exist.
b) The **speed of light** has the **same value** for all observers.

The Speed of Light and Relativity

The **invariance** of the speed of light is one of the cornerstones of special relativity. The other is based on the concept of an **inertial frame of reference**.

Anything Moving with a **Constant Velocity** is in an **Inertial Frame**

A reference frame is just a **space** that we decide to use to describe the **position of an object** — you can think of a reference frame as a **set of coordinates**.

An **inertial reference frame** is one in which **Newton's 1st law** is obeyed. (Newton's 1st law says that objects won't accelerate unless they're acted on by an external force.)

1) Imagine sitting in a carriage of a train **waiting at a station**. You put a **marble** on the table. The marble **doesn't move**, since there aren't any horizontal **forces** acting on it. **Newton's 1st law** applies, so it's an **inertial frame**.
2) You'll get the **same result** if the carriage moves at a **steady speed** (as long as the track is **smooth, straight and level**) — another inertial frame.
3) As the train **accelerates** out of the station, the marble **moves** without any force being applied. Newton's 1st law **doesn't apply**. The accelerating carriage **isn't an inertial frame**.
4) **Rotating** or **accelerating** reference frames **aren't** inertial. In most cases, though, you can think of the **Earth** as an inertial frame — it's near enough.

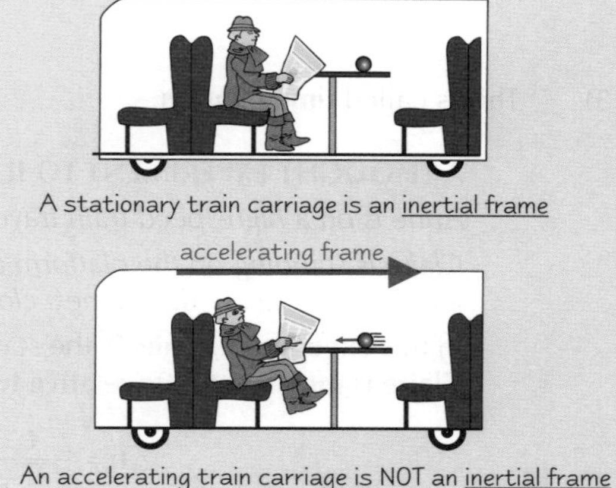

Einstein's **Postulates** of **Special Relativity**

Einstein's theory of **special relativity** only works in **inertial frames** and is based on **two postulates** (assumptions):

1) **Physical laws have the same form in all inertial frames.**
2) **The speed of light in free space is invariant.**

1) The first postulate says that if we do **any physics experiment** in any inertial frame we'll always get the **same result**. That means it's **impossible** to use the result of **any experiment** to work out if you're in a **stationary reference frame** or one moving at a **constant velocity**.
2) The second postulate says that the **speed of light** (in a vacuum) always has the **same value**. It isn't affected by the **movement** of the **person measuring it** or by the movement of the **light source**.

Practice Questions

Q1 Draw a labelled diagram showing the apparatus used by Michelson and Morley to determine the absolute speed of the Earth. Include the light source, mirrors, partial reflector, glass block and the position of the observer.

Q2 State the postulates of Einstein's theory of special relativity.

Q3 Explain why a carriage on a rotating Ferris wheel is not an inertial frame.

Exam Questions

Q1 In the Michelson-Morley interferometer experiment, interference fringes were observed. When the apparatus was rotated through 90 degrees the expected result was not observed.
(a) State the result that was expected. [1 mark]
(b) Describe the conclusions that were eventually drawn from these observations. [2 marks]

Q2 (a) Using a suitable example, explain what is meant by an inertial reference frame. [2 marks]
(b) Explain what is meant by the invariance of the speed of light. [2 marks]

The speed of light is always the same — whatever your reference frame...

Michelson and Morley showed that Newton's laws didn't always work. This was a huge deal. Newton's laws of motion had been treated like gospel by the physics community since the 17th century. Then along came Herr Einstein...

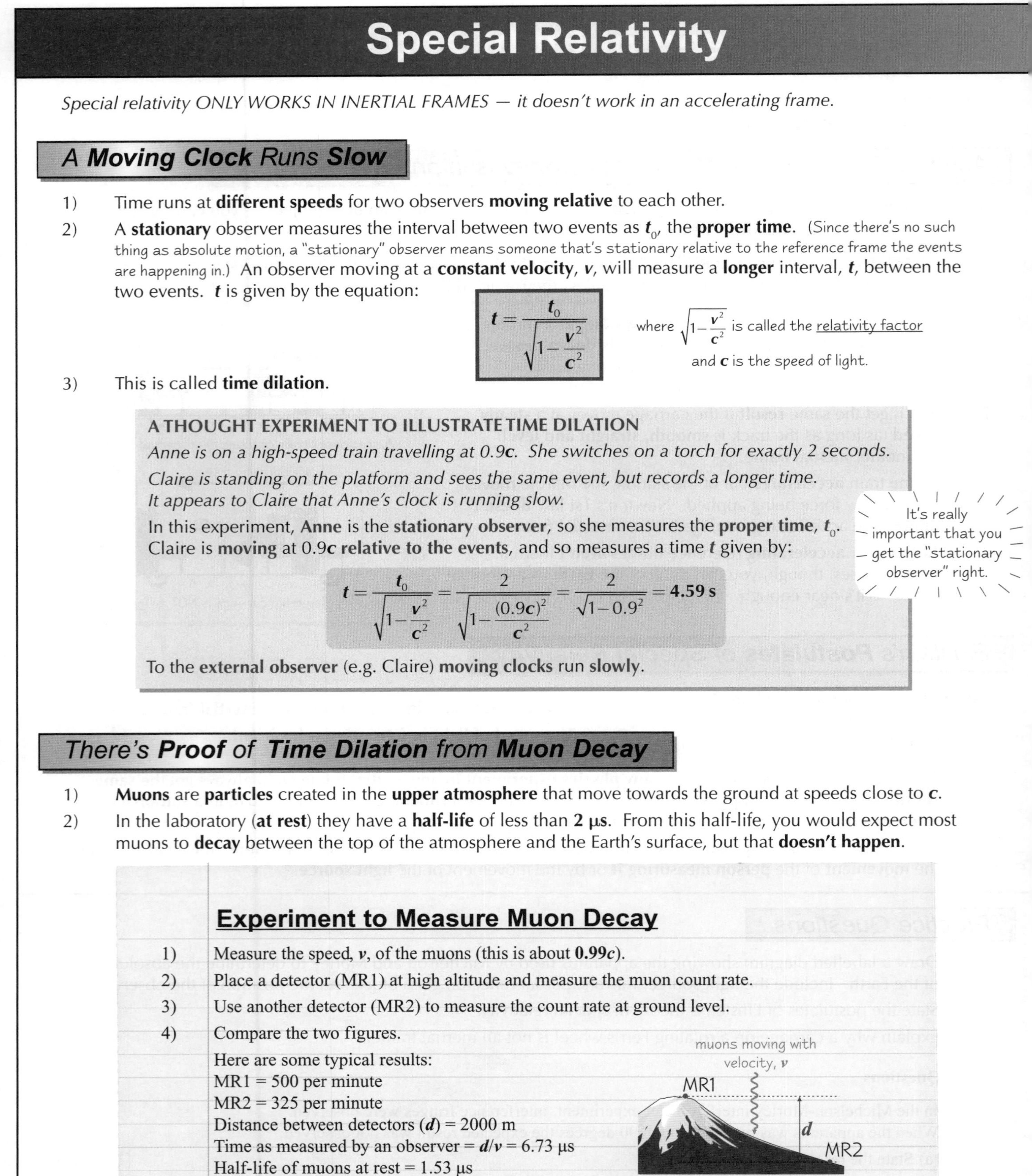

Special Relativity

Special relativity ONLY WORKS IN INERTIAL FRAMES — it doesn't work in an accelerating frame.

A Moving Clock Runs Slow

1) Time runs at **different speeds** for two observers **moving relative** to each other.
2) A **stationary** observer measures the interval between two events as t_0, the **proper time**. (Since there's no such thing as absolute motion, a "stationary" observer means someone that's stationary relative to the reference frame the events are happening in.) An observer moving at a **constant velocity**, v, will measure a **longer** interval, t, between the two events. t is given by the equation:

$$t = \frac{t_0}{\sqrt{1-\frac{v^2}{c^2}}}$$

where $\sqrt{1-\frac{v^2}{c^2}}$ is called the relativity factor and c is the speed of light.

3) This is called **time dilation**.

A THOUGHT EXPERIMENT TO ILLUSTRATE TIME DILATION

Anne is on a high-speed train travelling at 0.9c. She switches on a torch for exactly 2 seconds.
Claire is standing on the platform and sees the same event, but records a longer time.
It appears to Claire that Anne's clock is running slow.

In this experiment, **Anne** is the **stationary observer**, so she measures the **proper time**, t_0.
Claire is **moving** at 0.9**c relative to the events**, and so measures a time t given by:

$$t = \frac{t_0}{\sqrt{1-\frac{v^2}{c^2}}} = \frac{2}{\sqrt{1-\frac{(0.9c)^2}{c^2}}} = \frac{2}{\sqrt{1-0.9^2}} = \mathbf{4.59\ s}$$

To the **external observer** (e.g. Claire) **moving clocks** run **slowly**.

It's really important that you get the "stationary observer" right.

There's Proof of Time Dilation from Muon Decay

1) **Muons** are **particles** created in the **upper atmosphere** that move towards the ground at speeds close to **c**.
2) In the laboratory (**at rest**) they have a **half-life** of less than **2 μs**. From this half-life, you would expect most muons to **decay** between the top of the atmosphere and the Earth's surface, but that **doesn't happen**.

Experiment to Measure Muon Decay

1) Measure the speed, v, of the muons (this is about **0.99c**).
2) Place a detector (MR1) at high altitude and measure the muon count rate.
3) Use another detector (MR2) to measure the count rate at ground level.
4) Compare the two figures.

Here are some typical results:
MR1 = 500 per minute
MR2 = 325 per minute
Distance between detectors (**d**) = 2000 m
Time as measured by an observer = d/v = 6.73 μs
Half-life of muons at rest = 1.53 μs

muons moving with velocity, v
MR1
d
MR2

3) We can do some calculations using the data above. In the reference frame of the **observer** the muons seemed to have travelled for **4.4 half-lives** between the two detectors. You would expect the count rate at the **second detector** to be only about **25 counts per minute**.
4) However, in a **muon's reference frame**, travelling at 0.99c, the time taken for the journey is just t_0 **= 0.94 μs**. From the point of view of the muons, the time elapsed is **less** than their **half-life**. From the point of view of the **observer**, it appears that the half-life of the muons has been **extended**.

Special Relativity

A Moving Rod Looks Shorter

1) A **rod** moving in the **same direction** as its **length** looks **shorter** to an external observer.
2) A **stationary** observer measures the length of an object as l_0. An observer moving at a **constant velocity**, v, will measure a **shorter** length, l. l is given by the equation:

$$l = l_0\sqrt{1-\frac{v^2}{c^2}}$$

This is called **length contraction**.

A THOUGHT EXPERIMENT TO ILLUSTRATE LENGTH CONTRACTION

*Anne (still in the train moving at 0.9**c**) measures the length of her carriage as 3 m. Claire, on the platform, measures the length of the carriage as it moves past her.*

Claire measures a length: $l = l_0\sqrt{1-\frac{v^2}{c^2}} = 3\sqrt{1-\frac{(0.9c)^2}{c^2}} = 3\sqrt{1-0.9^2} = \mathbf{1.3\,m}$

The Mass of an Object Increases with Speed

1) The **faster** an object **moves**, the **more massive** it gets.
2) An object with rest mass m_0 moving at a **velocity** v has a **relativistic mass** m given by the equation:

$$m = \frac{m_0}{\sqrt{1-\frac{v^2}{c^2}}}$$

So increasing an object's kinetic energy increases its mass — but it's only noticeable near the speed of light.

3) As the relative speed of an object approaches **c**, the mass approaches **infinity**. So, in practice, no massive object can move at a speed **greater than** or **equal to** the speed of light.

Mass and Energy are Equivalent

1) Einstein extended his idea of **relativistic mass** to write down the most famous equation in physics: $E = mc^2$
2) This equation says that **mass** can be **converted** into **energy** and vice versa. Or, alternatively, **any energy** you supply to an object **increases** its **mass** — it's just that the increase is usually **too small** to measure.
3) The **total energy** of a relativistic object is given by the equation:

$$E = \frac{m_0c^2}{\sqrt{1-\frac{v^2}{c^2}}}$$

This is just substituting the relativistic mass into $E = mc^2$.

Practice Questions

Q1 State the equations for time dilation and length contraction, carefully defining each symbol.

Q2 Using the results from the muon experiment (page 100), show that the time elapsed in the reference frame of the muon is 0.94 μs.

Q3 A particle accelerated to near the speed of light gains a very large quantity of energy. Describe how the following quantities change as the particle gains more and more energy: a) the mass; b) the speed.

Exam Questions

Q1 A subatomic particle has a half-life of 20 ns when at rest. If a beam of these particles is moving at 0.995c relative to an observer, calculate the half-life of these particles in the frame of reference of the observer. [3 marks]

Q2 Describe a thought experiment to illustrate time dilation. [4 marks]

Q3 For a proton ($m_0 = 1.67 \times 10^{-27}$ kg) travelling at 2.8×10^8 ms^{-1} calculate:

(a) the relativistic mass, [1 mark]

(b) the total energy. [1 mark]

Have you ever noticed how time dilates when you're revising physics...

*In a moving frame, time stretches out, lengths get shorter and masses get bigger. One of the trickiest bits is remembering which observer's which — t_0, m_0 and l_0 are the values you'd measure if the object was **at rest**.*

Exponentials and Natural Logs

Mwah ha ha ha... you've hacked your way through the rest of the book and think you've finally got to the end of A2 Physics, but no, there's this tasty titbit of exam fun to go. You can get asked to look at and work out values from log graphs all over the shop, from astrophysics to electric field strength. And it's easy when you know how...

Many Relationships in Physics are **Exponential**

A fair few of the relationships you need to know about in A2 Physics are **exponential** — where the **rate of change** of a quantity is **proportional** to the **amount** of the quantity left. Here are just a few you should have met before (if they don't ring a bell, go have a quick read about them)...

Charge on a capacitor — the decay of charge on a capacitor is proportional to the amount of charge left on the capacitor: $Q = Q_0 e^{(-t/RC)}$ (see p 25)

Radioactive decay — the rate of decay is proportional to the **number of nuclei left** to decay in a sample: $N = N_0 e^{(-\lambda t)}$ (see p 39)

The **activity** of a radioactive sample behaves in the same way: $A = A_0 e^{(-\lambda t)}$ (see p 39)

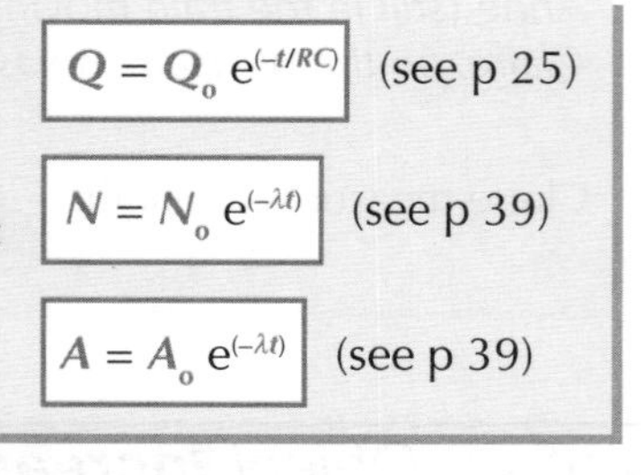

You can **Plot** Exponential Relations Using the **Natural Log, ln**

1) Say you've got two variables, x and y, which are related to each other by the formula $y = ke^{-ax}$ (where k and a are constants).
2) The inverse of e is the natural logarithm, **ln**.
3) By definition, $\ln(e^x) = x$. So far so good... now you need some **log rules**:

$\ln(ab) = \ln a + \ln b$ $\qquad \ln\left(\frac{a}{b}\right) = \ln a - \ln b$ $\qquad \ln a^b = b \ln a$

When it came to logs, Geoff always took time to smell the flowers...

4) So, if you take the natural log of the exponential function you get:
   $\ln y = \ln(ke^{-ax}) = \ln k + \ln(e^{-ax}) \Longrightarrow$ $\ln y = \ln k - ax$
5) Then all you need to do is plot ($\ln y$) against x, and Eric's your aunty:

   You get a **straight-line** graph with (**ln** k) as the **y-intercept**, and $-a$ as the **gradient**.

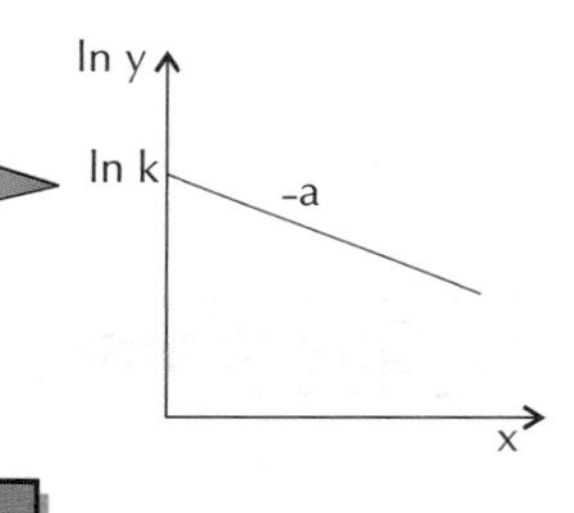

You Might be Asked to find the **Gradient** of a Log Graph...

This log business isn't too bad when you get your head around which bit of the log graph means what. On the plus side, they won't ask you to plot a graph like this (yipee) — they'll just want you to find the **gradient** or the **y-intercept**.

Example — finding the radioactive half-life of material X

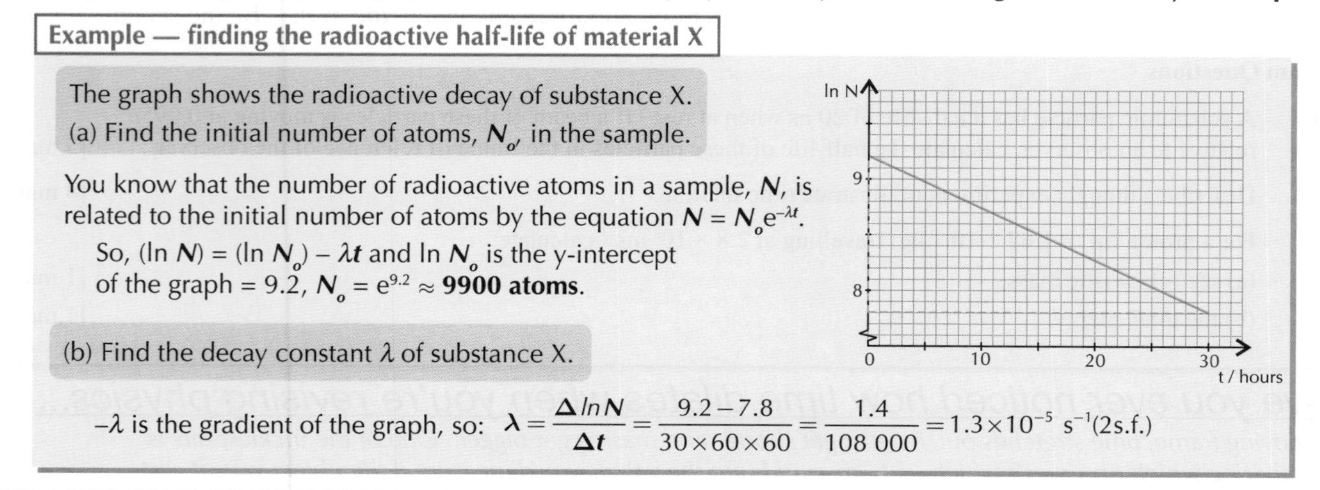

The graph shows the radioactive decay of substance X.
(a) Find the initial number of atoms, N_0, in the sample.

You know that the number of radioactive atoms in a sample, N, is related to the initial number of atoms by the equation $N = N_0 e^{-\lambda t}$.
So, $(\ln N) = (\ln N_0) - \lambda t$ and $\ln N_0$ is the y-intercept of the graph = 9.2, $N_0 = e^{9.2} \approx$ **9900 atoms**.

(b) Find the decay constant λ of substance X.

$-\lambda$ is the gradient of the graph, so: $\lambda = \frac{\Delta \ln N}{\Delta t} = \frac{9.2 - 7.8}{30 \times 60 \times 60} = \frac{1.4}{108\,000} = 1.3 \times 10^{-5}\ s^{-1}$ (2s.f.)

Log Graphs and Long Answer Questions

You can Plot **Any Power Law** as a **Log-Log Graph**

You can use logs to plot a straight-line graph of **any power law** — it doesn't have to be an exponential.
Take the relationship between the energy stored in a spring, E, and the spring's extension, x:

$$E = kx^n$$

Take the log (base 10) of both sides to get:

$$\log E = \log k + n \log x$$

So $\log k$ will be the y-intercept and n the gradient of the graph.

Example

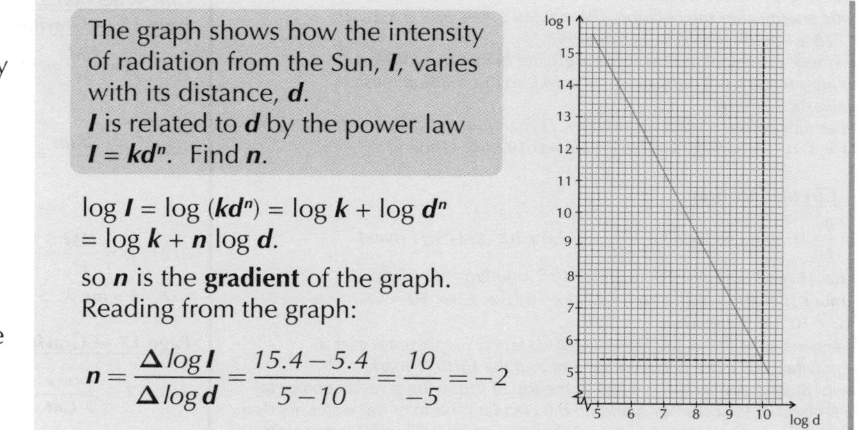

The graph shows how the intensity of radiation from the Sun, I, varies with its distance, d.
I is related to d by the power law $I = kd^n$. Find n.

$\log I = \log (kd^n) = \log k + \log d^n$
$= \log k + n \log d$.

so n is the **gradient** of the graph.
Reading from the graph:

$$n = \frac{\Delta \log I}{\Delta \log d} = \frac{15.4 - 5.4}{5 - 10} = \frac{10}{-5} = -2$$

And that's the End of Logs... Now **Explain Yourself...**

In A2, they often give a couple of marks for 'the quality of written communication' when you're writing a slightly long answer (and not just pumping numbers into an equation). You can pick up a couple of easy marks just by making sure that you do the things in the fetching blue box.

1) **Explain** your ideas or argument **clearly** as this is usually what you'll get a mark for. And make sure you **answer the question** being asked — it's dead easy to go off on a tangent. Like my mate Phil always says... have I ever told you about Phil? Well he...
2) Write in **whole sentences**.
3) Use **correct spelling**, **grammar** and **punctuation**.
4) Also check how many marks the question is worth. If it's only a two-marker, they don't want you to spend half an hour writing an essay about it.

Example

Human hearing is most sensitive around the resonant frequency of the ear passages — around 3000 Hz. Explain fully what is meant by this statement and describe two circumstances in which you would expect a **higher** resonant frequency in a human.
The quality of your written answer will be assessed in this question. *[6 marks]*

Good Answer

Resonance happens when the driving frequency is the same as the natural frequency of the oscillator.
The amplitude of the oscillations will be greater at the resonant frequency for any given amplitude of driving wave.
The resonant frequency of the ear passages depends on their length. In a child, for example, the ear passages are shorter than average, so the resonant frequency is higher.
You would also expect the ear to have a higher resonant frequency if the auditory canal was filled with e.g. water rather than air. Sound travels faster through water than air, so the resonant frequency of the canal would be higher.

Bad Answer

driving frequency = nat. freq.
shorter ear passages = higher res. freq.
length varies between humans.
sound faster in water.

There's nothing wrong with the physics in the bad answer, but you'd miss out on some nice easy marks just for not bothering link you thoughts together properly or put your answer into proper sentences.

Lumberjacks are great musicians — they have a natural logarithm...

Well, that's it folks. Crack open the chocolate bar of victory and know you've earnt it. Only the tiny detail of the actual exam to go... ahem. Make sure you know which bit means what on a log graph and you'll pick up some nice easy marks. Other than that, stay calm, be as clear as you can and good luck — I've got my fingers, toes and eyes crossed for you.

Answers

Unit 4: Section 1 — Further Mechanics

Page 5 — Momentum and Impulse

1) a) total momentum before collision = total momentum after [1 mark]
$(0.6 \times 5) + 0 = (0.6 \times -2.4) + 2\mathbf{v}$
$3 + 1.44 = 2\mathbf{v}$ [1 mark for working] $\Rightarrow \mathbf{v} = 2.22$ ms^{-1} [1 mark]

b) Kinetic energy before collision = $\frac{1}{2} \times 0.6 \times 5^2 + \frac{1}{2} \times 2 \times 0^2 = 7.5$ J
Kinetic energy after the collision = $\frac{1}{2} \times 0.6 \times 2.4^2 + \frac{1}{2} \times 2 \times 2.22^2$
$= 1.728 + 4.9284 = 6.6564$ J [1 mark]
The kinetic energy of the two balls is greater before the collision than after (i.e. it's not conserved) [1 mark], so the collision must be inelastic [1 mark].

2) momentum before = momentum after [1 mark] $\Rightarrow (0.7 \times 0.3) + 0 = 1.1\mathbf{v}$
$0.21 = 1.1\mathbf{v}$ [1 mark for working] $\Rightarrow \mathbf{v} = 0.19$ ms^{-1} [1 mark]

Page 7 — Circular Motion

1) a) $\omega = \frac{\theta}{t}$ [1 mark] so $\omega = \frac{2\pi}{3.2 \times 10^7} = 2.0 \times 10^{-7}$ rad s^{-1} [1 mark]

b) $\mathbf{v} = \mathbf{r}\omega$ [1 mark] $= 1.5 \times 10^{11} \times 2.0 \times 10^{-7} = 30$ kms^{-1} [1 mark]

c) $\mathbf{F} = \mathbf{m}\omega^2\mathbf{r}$ [1 mark] $= 6.0 \times 10^{24} \times (2.0 \times 10^{-7})^2 \times 1.5 \times 10^{11}$
$= 3.6 \times 10^{22}$ N [1 mark]
The answers to b) and c) use the rounded value of ω calculated in part a).

d) The gravitational force between the Sun and the Earth [1 mark]

2) a) Gravity pulling down on the water at the top of the swing gives a centripetal acceleration of 9.81 ms^{-2} [1 mark]. If the circular motion of the water needs a centripetal acceleration of less than 9.81 ms^{-2}, gravity will pull it in too tight a circle. The water will fall out of the bucket.

Since $\mathbf{a} = \omega^2\mathbf{r}$, $\omega^2 = \frac{\mathbf{a}}{\mathbf{r}} = \frac{9.81}{1}$, so $\omega = 3.1$ rad s^{-1} [1 mark]

$\omega = 2\pi\mathbf{f}$, so $\mathbf{f} = \frac{\omega}{2\pi} = 0.5$ rev s^{-1} [1 mark]

b) Centripetal force $= \mathbf{m}\omega^2\mathbf{r} = 10 \times 5^2 \times 1 = 250$ N [1 mark].
This force is provided by both the tension in the rope, $\mathbf{T}$, and gravity:
$\mathbf{T} + (10 \times 9.81) = 250$. So $\mathbf{T} = 250 - (10 \times 9.81) = 152$ N [1 mark].

Page 9 — Simple Harmonic Motion

1) a) Simple harmonic motion is an oscillation in which an object always accelerates towards a fixed point [1 mark] with an acceleration directly proportional to its displacement from that point [1 mark]. [The SHM equation would get you the marks if you defined all the variables.]

b) The acceleration of a falling bouncy ball is due to gravity. This acceleration is constant, so the motion is not SHM. [1 mark].

2) a) Maximum velocity $= (2\pi\mathbf{f})\mathbf{A} = 2\pi \times 1.5 \times 0.05 = 0.47$ ms^{-1} [1 mark].

b) Stopclock started when object released, so $\mathbf{x} = \mathbf{A}\cos(2\pi\mathbf{f}\mathbf{t})$ [1 mark].
$\mathbf{x} = 0.05 \times \cos(2\pi \times 1.5 \times 0.1) = 0.05 \times \cos(0.94) = 0.029$ m [1 mark].

c) $\mathbf{x} = \mathbf{A}\cos(2\pi\mathbf{f}\mathbf{t}) \Rightarrow 0.01 = 0.05 \times \cos(2\pi \times 1.5\mathbf{t})$.
So $0.2 = \cos(3\pi\mathbf{t}) \Rightarrow \cos^{-1}(0.2) = 3\pi\mathbf{t}$. $3\pi\mathbf{t} = 1.37 \Rightarrow \mathbf{t} = 0.15$ s.
[1 mark for working, 1 mark for correct answer]

Page 11 — Simple Harmonic Oscillators

1) a) Extension of spring = 0.20 – 0.10 = 0.10 m [1 mark]. Hooke's Law gives

$\mathbf{k} = \frac{\text{force}}{\text{extension}}$, so $\mathbf{k} = \frac{0.10 \times 9.8}{0.10} = 9.8$ Nm^{-1} [1 mark].

b) $\mathbf{T} = 2\pi\sqrt{\frac{\mathbf{m}}{\mathbf{k}}} \Rightarrow \mathbf{T} = 2\pi \times \sqrt{\frac{0.10}{9.8}} = 2\pi \times \sqrt{0.01} = 0.63$ s [1 mark].

c) $\mathbf{m} \propto \mathbf{T}^2$ so if $\mathbf{T}$ is doubled, $\mathbf{T}^2$ is quadrupled and $\mathbf{m}$ is quadrupled [1 mark].
So mass needed = 4 × 0.10 = 0.40 kg [1 mark].

2) E.g. $5\mathbf{T}_{\text{short pendulum}} = 3\mathbf{T}_{\text{long pendulum}}$, and $\mathbf{T} = 2\pi\sqrt{\frac{\mathbf{l}}{\mathbf{g}}}$ [1 mark]. Let length of long pendulum = $\mathbf{l}$. So $5\left(2\pi\sqrt{\frac{0.20}{\mathbf{g}}}\right) = 3\left(2\pi\sqrt{\frac{\mathbf{l}}{\mathbf{g}}}\right)$ [1 mark]. Dividing by 2π gives $5 \times \sqrt{\frac{0.20}{\mathbf{g}}} = 3 \times \sqrt{\frac{\mathbf{l}}{\mathbf{g}}}$. Squaring and simplifying gives $5 = 9\mathbf{l}$ so length of long pendulum = 5/9 = 0.56 m [1 mark].

Page 13 — Free and Forced Vibrations

1) a) When a system is forced to vibrate at a frequency that's close to, or the same as its natural frequency [1 mark] and oscillates with a much larger than usual amplitude [1 mark].

b) [1 mark] for a peak at the natural frequency, [1 mark] for a sharp peak.

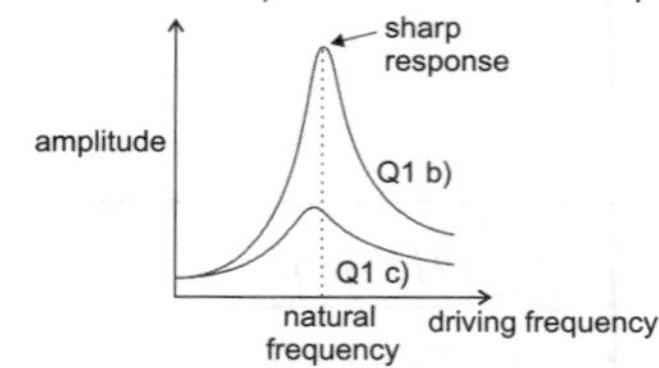

c) See graph. [1 mark] for a smaller peak at the natural frequency [the peak will actually be slightly to the left of the natural frequency, but you'll get the mark if the peak is at the same frequency in the diagram].

2) a) A system is critically damped if it returns to rest in the shortest time possible [1 mark] when it's displaced from equilibrium and released.

b) e.g. suspension in a car [1 mark].

Unit 4: Section 2 — Gravitation and Electric Fields

Page 15 — Gravitational Fields

1) a) $\mathbf{g} = \frac{\mathbf{GM}}{\mathbf{r}^2} \Rightarrow \mathbf{M} = \frac{\mathbf{gr}^2}{\mathbf{G}} = \frac{9.81 \times (6400 \times 1000)^2}{6.67 \times 10^{-11}}$ [1 mark]
$= 6.02 \times 10^{24}$ kg [1 mark]

b) $\mathbf{F} = \frac{\mathbf{GMm}}{\mathbf{r}^2} = \frac{6.67 \times 10^{-11} \times 1.99 \times 10^{30} \times 6.02 \times 10^{24}}{(1.5 \times 10^{11})^2}$ [1 mark]
$= 3.6 \times 10^{22}$ N [1 mark]

2) a) $\mathbf{g} = \frac{\mathbf{GM}}{\mathbf{r}^2} = \frac{6.67 \times 10^{-11} \times 7.35 \times 10^{22}}{(1740 \times 1000)^2} = 1.62$ Nkg^{-1} [1 mark]

b) $\mathbf{F} = \mathbf{mg} = 25 \times 1.62 = 40.5$ N [1 mark]

Page 17 — Gravitational Fields

1) a) $\mathbf{T} = \sqrt{\frac{4\pi^2\mathbf{r}^3}{\mathbf{GM}}} = \sqrt{\frac{4\pi^2 \times [(6400 + 200) \times 1000]^3}{6.67 \times 10^{-11} \times 5.98 \times 10^{24}}}$ [1 mark]
= 5334 seconds OR 1.48 hours [1 mark]

b) $\mathbf{v} = \sqrt{\frac{\mathbf{GM}}{\mathbf{r}}} = \sqrt{\frac{6.67 \times 10^{-11} \times 5.98 \times 10^{24}}{(6400 + 200) \times 1000}}$ [1 mark]
$= 7774$ ms$^{-1} \approx 7.77$ kms^{-1}

2) Period = 24 hours = 24 × 60 × 60 = 86400 s [1 mark]

$\mathbf{T} = \sqrt{\frac{4\pi^2\mathbf{r}^3}{\mathbf{GM}}} \Rightarrow \mathbf{r} = \sqrt[3]{\frac{\mathbf{T}^2\mathbf{GM}}{4\pi^2}} = \sqrt[3]{\frac{86400^2 \times 6.67 \times 10^{-11} \times 5.98 \times 10^{24}}{4\pi^2}}$

$\mathbf{r} = 4.23 \times 10^7$ m $= 4.23 \times 10^4$ km [1 mark]
Height above Earth $= 4.23 \times 10^4 - 6.4 \times 10^3 = 35900$ km [1 mark]

Page 19 — Electric Fields

1) Charge on alpha particle, $\mathbf{Q}_1 = +2e = 2 \times 1.6 \times 10^{-19} = 3.2 \times 10^{-19}$ C
Charge on gold nucleus, $\mathbf{Q}_2 = +79e = 79 \times 1.6 \times 10^{-19}$ C
$= 1.3 \times 10^{-17}$ C [1 mark]

$\mathbf{F} = \frac{1}{4\pi\varepsilon_0}\frac{\mathbf{Q}_1\mathbf{Q}_2}{\mathbf{r}^2} = \frac{1}{4\pi\varepsilon_0}\frac{3.2 \times 10^{-19} \times 1.3 \times 10^{-17}}{(5 \times 10^{-12})^2}$ [1 mark]

$= 1.5 \times 10^{-3}$ N [1 mark] away from the gold nucleus [1 mark]

2) a) $\mathbf{E} = \mathbf{V}/\mathbf{d} = 1500/(4.5 \times 10^{-3}) = 3.3 \times 10^5$ [1 mark] Vm^{-1} [1 mark]
The field is perpendicular to the plates. [1 mark]

b) $\mathbf{d} = 2 \times (4.5 \times 10^{-3}) = 9.0 \times 10^{-3}$ m [1 mark]
$\mathbf{E} = \mathbf{V}/\mathbf{d} \Rightarrow \mathbf{V} = \mathbf{Ed} = 3.3 \times 10^5 \times 9 \times 10^{-3} = 3000$ V [1 mark]

Page 21 — Electric Fields

1) $\mathbf{F}_g = -\frac{\mathbf{Gm}_1\mathbf{m}_2}{\mathbf{r}^2} = \frac{6.67 \times 10^{-11} \times (9.11 \times 10^{-31})^2}{(8 \times 10^{-10})^2} = -8.65 \times 10^{-53}$ N

$\mathbf{F}_e = \frac{1}{4\pi\varepsilon_0}\frac{\mathbf{Q}_1\mathbf{Q}_2}{\mathbf{r}^2} = \frac{1}{4\pi\varepsilon_0}\frac{(1.60 \times 10^{-19})^2}{(8 \times 10^{-10})^2} = 3.60 \times 10^{-10}$ N [1 mark]

The electric force on each electron is much larger than the gravitational force, by a factor of over 10^{40} [1 mark]. The gravitational force is attractive, while the electric force is repulsive [1 mark].

2) a) Oil drop is stationary, so $\mathbf{mg} = \mathbf{F}_e = \mathbf{Vq}/\mathbf{d}$ [1 mark]
$\Rightarrow \mathbf{q} = \mathbf{mgd}/\mathbf{V} = (1.5 \times 10^{-14} \times 9.81 \times 0.03)/5000$ [1 mark]
$= 8.8 \times 10^{-19}$ C [1 mark]

b) The drop would accelerate towards the positive lower plate [1 mark].

Unit 4: Section 3 — Capacitance

Page 23 — Capacitors

1) a) Capacitance is the amount of energy stored per volt. [1 mark]

b)i) Capacitance $= \frac{\mathbf{Q}}{\mathbf{V}}$ = gradient of line $= \frac{660\ \mu C}{3\ V} = 220\ \mu F$.
[1 mark for 'gradient', 1 mark for correct answer.]

ii) Charge stored $= \mathbf{Q}$ = area $= 15 \times 10^{-6} \times 66 = 990\ \mu C$.
[1 mark for 'area', 1 mark for correct answer.]

2) a) E.g.

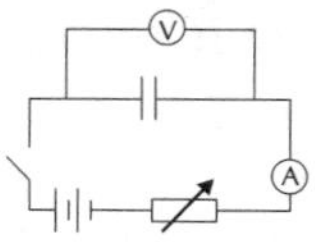

[2 marks, or 1 mark for three components correct]

Answers

b) Close the switch and start a stopwatch. Adjust the variable resistor to keep the current constant as the capacitor charges [1 mark]. Record the p.d. across the capacitor at regular intervals until it equals the battery voltage [1 mark]. Plot a graph of charge stored (using $Q = It$) against voltage, then find the area under it. This is the energy stored by the capacitor [1 mark].

3) a) $E = \frac{1}{2}CV^2$ [1 mark] $= \frac{1}{2} \times 0.5 \times 12^2 = 36$ J [1 mark]

b) $Q = CV$ [1 mark] $= 0.5 \times 12 = 6$ C [1 mark]

Page 25 — Charging and Discharging

1) a) The charge falls to 37% after RC seconds [1 mark], so $t = 1000 \times 2.5 \times 10^{-4} = 0.25$ seconds [1 mark]

b) $Q = Q_0 e^{-\frac{t}{RC}}$ [1 mark], so after 0.7 seconds: $Q = Q_0 e^{-\frac{0.7}{0.25}} = Q_0 \times 0.06$ [1 mark]. 6% of the initial charge is left on the capacitor after 0.7 s [1 mark].

c) i) The total charge stored doubles: V is proportional to Q [1 mark].
ii) None: this is a fixed property of the capacitor [1 mark].
iii) None: charging time depends only on the capacitance of the capacitor and the resistance of the circuit, which don't change [1 mark].

Unit 4: Section 4 — Magnetic Fields

Page 27 — Magnetic Fields

1) a) $F = BIl = 2 \times 10^{-5} \times 3 \times 0.04$ [1 mark] $= 2.4 \times 10^{-6}$ N [1 mark]

b) The force is zero [1 mark] because the two magnetic fields are perpendicular, so do not affect each other [1 mark].

Page 29 — Electromagnetic Induction

1) a) $\phi = BA$ [1 mark] $= 2 \times 10^{-3} \times 0.23 = 4.6 \times 10^{-4}$ Wb [1 mark]

b) $\Phi = BAN$ [1 mark] $= 2 \times 10^{-3} \times 0.23 \times 150 = 0.069$ Wb [1 mark]

c) $$V = \frac{\Delta\Phi}{\Delta t} = \frac{(B_{start} - B_{end})AN}{\Delta t} = \frac{(2\times10^{-3} - 1.5\times10^{-3})(0.23\times150)}{2.5} = 6.9\times10^{-3}\text{ V}$$

[3 marks available, one for each stage of the workings]

2) a) $V = Blv$ [1 mark] $= 60 \times 10^{-6} \times 30 \times 100 = 0.18$ V [1 mark]

b) 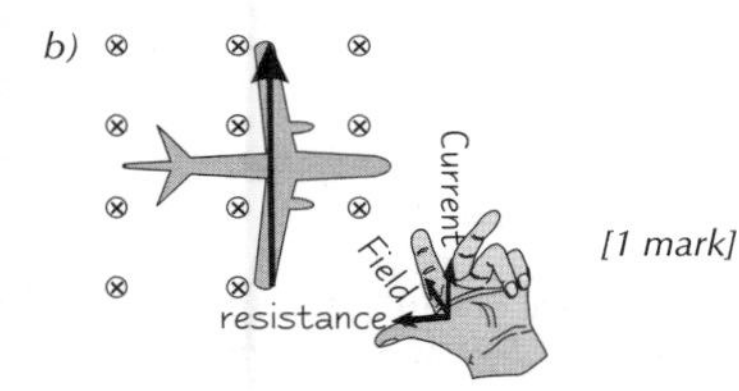

[1 mark]

Page 31 — Transformers and Alternators

1) a) $\frac{V_p}{V_s} = \frac{N_p}{N_s}$ [1 mark] so, $N_s = \frac{45\times150}{9} = 750$ turns [1 mark]

b) $\frac{V_p}{V_s} = \frac{I_s}{I_p}$ [1 mark] so, $I_s = \frac{V_p I_p}{V_s} = \frac{9\times1.5}{45} = 0.3$ A [1 mark]

c) efficiency $= \frac{V_s I_s}{V_p I_p}$ [1 mark] $= \frac{10.8}{9\times1.5} = 0.8$ (i.e. 80%) [1 mark]

2) a) $\Phi = BAN \cos\theta$ [1 mark] $= 0.9 \times 0.01 \times 500 \times \cos 60° = 2.25$ Wb [1 mark]

b) Peak e.m.f. when $\sin \omega t = \pm1$, giving: $\varepsilon = \pm BAN\omega$ [1 mark]
So, peak e.m.f. is: $\varepsilon = \pm 0.9 \times 0.01 \times 500 \times 40\pi$ [1 mark] $\varepsilon = \pm565$ V [1 mark]

Unit 5: Section 1 — Radioactivity and Nuclear Energy

Page 33 — Scattering to Determine Structure

1) a) The majority of alpha particles are not scattered because the nucleus is a very small part of the whole atom and so the probability of an alpha particle getting near it is small [1 mark]. Most alpha particles pass undeflected through the empty space around the nucleus [1 mark].

b) Alpha particles and atomic nuclei are both positively charged [1 mark]. If an alpha particle travels close to a nucleus, there will be a significant electrostatic force of repulsion between them [1 mark]. This force deflects the alpha particle from its original path. [1 mark]

2) a) All particles have wave-like properties, with an associated wavelength [1 mark]. If the wavelength of a beam of particles is similar to the atomic spacing of the material it's passing through, the beam will produce a diffraction pattern. [1 mark]

b) Electrons are not affected by the strong nuclear force. [1 mark]

c) Maximum diffraction occurs when the nucleus is the same size as the wavelength of the electrons [1 mark]. Larger nuclei cause less diffraction for the same electron energy [1 mark].

Page 35 — Nuclear Radius and Density

1) a) Rearrange $r = r_0 A^{1/3}$ [1 mark], then substitute the given values of r and A:

$$r_0 = \frac{r}{A^{1/3}} = \frac{3.2\times10^{-15}}{12^{1/3}} = 1.40\times10^{-15}\text{ m}$$ [1 mark]

b) For a radium nucleus with $A = 226$:

$r = r_0 A^{1/3} = 1.4\times10^{-15} \times 226^{1/3} = 8.53\times10^{-15}$ m [1 mark]

c) Volume $= \frac{4}{3}\pi r^3 = \frac{4}{3}\pi(8.53\times10^{-15})^3 = 2.6\times10^{-42}$ m^3 [1 mark]

So density $(\rho) = \frac{m}{V} = \frac{3.75\times10^{-25}}{2.6\times10^{-42}} = 1.44\times10^{17}$ kgm^{-3} [1 mark]

2) The mass density of a gold nucleus is much larger than the mass density of a gold atom [1 mark]. This implies that the majority of a gold atom's mass is contained in the nucleus [1 mark].
The nucleus is small compared to the size of the atom [1 mark].
There must be a lot of nearly empty space inside each atom [1 mark].

Page 37 — Radioactive Emissions

1) Place different materials between the source and detector and measure the amount of radiation getting through [1 mark]:

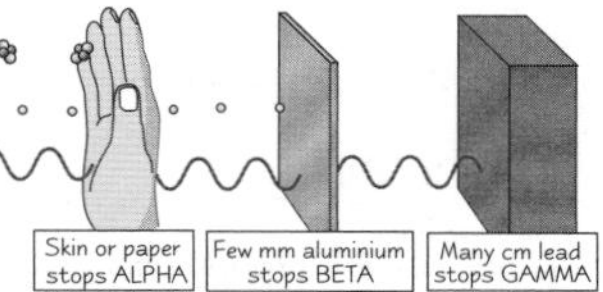

[1 mark for each material stopping correct radiation]

2) $I \propto$ count rate $\propto \frac{1}{x^2}$ [1 mark]

The G-M tube is 4 times the original distance from the source.

$\frac{1}{4^2} = \frac{1}{16}$, so the count rate at 40 cm will be 1/16th that at 10 cm [1 mark].

The count rate at 40 cm will be $\frac{240}{16} = 15$ counts s^{-1} [1 mark].

Page 39 — Exponential Law of Decay

1) Any one of: You can't say which atom/nucleus in a sample will decay next. / You can only estimate the fraction of nuclei that will decay or the probability an atom/nucleus will decay in a given time. / You cannot say exactly how many atoms will decay in a given time. [1 mark]

2) a) Activity, A = measured activity – background activity = 750 – 50 = 700 Bq [1 mark]
$A = \lambda N \Rightarrow 700 = 50\,000\,\lambda$ [1 mark] So $\lambda = 0.014$ s^{-1} [1 mark]

b) $T_{\frac{1}{2}} = \frac{\ln 2}{\lambda} = \frac{0.693}{0.014} = 49.5$ seconds

[1 mark for half-life equation, 1 mark for correct half-life]

c) $N = N_0 e^{-\lambda t} = 50\,000 \times e^{-0.014 \times 300} = 750$
[1 mark for decay equation, 1 mark for correct number of atoms]

Page 41 — Nuclear Decay

1) a) ${}^{226}_{88}Ra \rightarrow {}^{222}_{86}Rn + {}^{4}_{2}\alpha$ [3 marks available — 1 mark for alpha particle, 1 mark each for proton and nucleon number of radon]

b) ${}^{40}_{19}K \rightarrow {}^{40}_{20}Ca + {}^{0}_{-1}\beta$ [3 marks available — 1 mark for beta particle, 1 mark each for proton and nucleon number of calcium]

2) Mass defect $= (6.695 \times 10^{-27}) - (6.645 \times 10^{-27}) = 5.0 \times 10^{-29}$ kg [1 mark].
Using the equation $E = mc^2$ [1 mark],
$E = (5.0 \times 10^{-29}) \times (3 \times 10^8)^2 = 4.5 \times 10^{-12}$ J [1 mark]

Page 43 — Binding Energy

1) a) There are 6 protons and 8 neutrons, so the mass of individual parts $= (6 \times 1.007276) + (8 \times 1.008665) = 14.112976$ u [1 mark]
Mass of ${}^{14}_{6}C$ nucleus = 13.999948 u
so, mass defect = 14.112976 – 13.999948 = 0.113028 u [1 mark]
Converting this into kg gives mass defect $= 1.88 \times 10^{-28}$ kg [1 mark]

b) $E = mc^2 = (1.88 \times 10^{-28}) \times (3 \times 10^8)^2 = 1.69 \times 10^{-11}$ J [1 mark]
1 MeV $= 1.6 \times 10^{-13}$ J, so energy $= \frac{1.69\times10^{-11}}{1.6\times10^{-13}} = 106$ MeV [1 mark]

2) a) Fusion [1 mark]

b) The increase in binding energy per nucleon is about 0.86 MeV [1 mark]. There are 2 nucleons in ^2H, so the increase in binding energy is about 1.72 MeV — so about 1.7 MeV is released (ignoring the positron) [1 mark].

Answers

Page 45 — Nuclear Fission and Fusion

1) a) Nuclear fission can be induced by neutrons and produces more neutrons during the process [1 mark]. This means that each fission reaction induces more fission reactions, resulting in a ongoing chain of reactions [1 mark].

b) For example, control rods limit the rate of fission by absorbing neutrons [1 mark]. The number of neutrons absorbed by the rods is controlled by varying the amount they are inserted into the reactor [1 mark]. A suitable material for the control rods is boron [1 mark].

c) In an emergency shut-down, the control rods are released into the reactor [1 mark]. The control rods absorb the neutrons, and stop the reaction as quickly as possible [1 mark].

2) Advantages (any two of e.g.): the nuclear reactor itself doesn't produce any waste gases that could be harmful to the environment e.g. sulfur dioxide (leading to acid rain) or carbon dioxide [1 mark]. It can be used to supply a continuous supply of electricity unlike some renewable sources [1 mark]. Disadvantages (any two of e.g.): problems with the reactor getting out of control [1 mark], risks of radiation from radioactive waste produced [1 mark], the long half-life of nuclear waste [1 mark].

Unit 5: Section 2 — Thermal Physics

Page 47 — Ideal Gases

1) a)i) Number of moles $= \frac{\text{mass of gas}}{\text{molar mass}} = \frac{0.014}{0.028} = 0.5$ [1 mark]

ii) Number of molecules = number of moles × Avogadro's constant $= 0.5 \times 6.02 \times 10^{23} = 3.01 \times 10^{23}$ [1 mark]

b) $\mathbf{pV} = \mathbf{nRT}$, so $\mathbf{p} = \frac{\mathbf{nRT}}{\mathbf{V}}$ [1 mark] $= \frac{0.5 \times 8.31 \times 300}{0.01} = 125\ 000$ Pa [1 mark]

c) The pressure would also halve [1 mark] because it is proportional to the number of molecules — $pV = NkT$ [1 mark].

2) At ground level, $\frac{\mathbf{pV}}{\mathbf{T}} = \frac{1 \times 10^5 \times 10}{293} = 3410\ \text{JK}^{-1}$ [1 mark]

$\mathbf{pV/T}$ is constant, so higher up $\mathbf{pV/T} = 3410\ \text{JK}^{-1}$ [1 mark]

Higher up, $\mathbf{p} = \frac{3410 \times \mathbf{T}}{\mathbf{V}} = \frac{3410 \times 260}{25} = 35\ 500$ Pa [1 mark]

Page 49 — The Pressure of an Ideal Gas

1) a) $\mathbf{pV} = \frac{1}{3}\mathbf{Nm}\overline{\mathbf{c}^2}$ Rearrange the equation:

$\overline{\mathbf{c}^2} = \frac{3\mathbf{pV}}{\mathbf{Nm}}$ [1 mark] $= \frac{3 \times 1 \times 10^5 \times 7 \times 10^{-5}}{2 \times 10^{22} \times 6.6 \times 10^{-27}} = 159\ 091\ (\text{ms}^{-1})^2$ [1 mark]

b) r.m.s. speed $= \sqrt{\overline{c^2}} = \sqrt{159\ 091} = 399\ \text{ms}^{-1}$ [1 mark]

c) pV is proportional to T, so doubling T will double pV. [1 mark]

r.m.s. speed $= \sqrt{\overline{c^2}} = \sqrt{\frac{3\mathbf{pV}}{\mathbf{Nm}}}$, so doubling pV will increase the r.m.s. speed by a factor of $\sqrt{2}$. r.m.s. speed $= 399 \times \sqrt{2} = 564\ \text{ms}^{-1}$ [1 mark]

Page 51 — Energy and Temperature

1) a) $\frac{1}{2}\mathbf{m}\overline{\mathbf{c}^2} = \frac{3\mathbf{kT}}{2}$ Rearranging gives: $\overline{\mathbf{c}^2} = \frac{3\mathbf{kT}}{\mathbf{m}}$ [1 mark]

$\mathbf{m}$ = mass of 1 mole ÷ $N_A = 2.8 \times 10^{-2} \div 6.02 \times 10^{23} = 4.65 \times 10^{-26}$ kg [1 mark]

$\overline{\mathbf{c}^2} = \frac{3 \times 1.38 \times 10^{-23} \times 300}{4.65 \times 10^{-26}} = 2.67 \times 10^5\ \text{m}^2\text{s}^{-2}$ [1 mark]

Typical speed = r.m.s. speed $= \sqrt{2.67 \times 10^5} = 517\ \text{ms}^{-1}$ [1 mark]

b) Gas molecules move at different speeds because they have different amounts of energy [1 mark]. The molecules have different amounts of energy because they constantly collide and transfer energy between themselves [1 mark].

2) a) Time = distance ÷ speed = 8.0 m ÷ 400 = 0.02 s [1 mark]

b) Although the particles move at 400 ms^{-1} on average, they frequently collide with other particles [1 mark]. So their motion in any one direction is limited and they only slowly move from one end of the room to the other [1 mark].

3) Electrical energy supplied: $\Delta\mathbf{Q} = \mathbf{VI}\Delta\mathbf{t} = 12 \times 7.5 \times 180 = 16200$ J [1 mark]

The temperature rise is 12.7 – 4.5 = 8.2 °C

$\mathbf{c} = \frac{\Delta \mathbf{Q}}{\mathbf{m}\Delta\theta}$ [1 mark] $= 16200 \div (2 \times 8.2) = 988\ \text{Jkg}^{-1}\text{°C}^{-1}$ [1 mark]

You need the right unit for the third mark — $\text{J kg}^{-1}\ \text{K}^{-1}$ would be right too.

4) $\Delta\mathbf{Q} = \mathbf{ml} = 2.26 \times 10^6 \times 0.5 = 1.13 \times 10^6$ J [1 mark]

3 kW means 3000 J in 1s, so time $= 1.13 \times 10^6 / 3000$ [1 mark] = 377 s [1 mark]

Unit 5: Option A — Astrophysics and Cosmology

Page 54 — Optical Telescopes

1) a) lens axis

principal axis

The principal focus is where rays parallel to the principal axis converge [1 mark]. The focal length is the distance between the lens axis and the principal focus [1 mark].

b) $\frac{1}{\mathbf{u}} + \frac{1}{\mathbf{v}} = \frac{1}{\mathbf{f}}$ [1 mark]. So $\frac{1}{0.2} + \frac{1}{\mathbf{v}} = \frac{1}{0.15}$ [1 mark].

$\frac{1}{\mathbf{v}} = \frac{1}{0.15} - \frac{1}{0.2} = \frac{5}{3} \Rightarrow \mathbf{v} = 0.6\text{m}$ [1 mark].

c) Using lens equation: $\frac{1}{\mathbf{v}} = \frac{1}{0.15} - \frac{1}{0.10} = -\frac{10}{3} \Rightarrow \mathbf{v} = -0.3\text{m}$ [1 mark]

The sign of **v** is negative, indicating that the image is a virtual image on the same side of the lens as the object [1 mark].

2) a) Separation of lenses needs to be $\mathbf{f_o} + \mathbf{f_e} = 5.0 + 0.10 = 5.1$ m [1 mark].

b) Angular magnification = angle subtended by image at eye / angle subtended by object at unaided eye [1 mark]. $\mathbf{M} = \mathbf{f_o}/\mathbf{f_e} = 5.0 / 0.10 = 50$ [1 mark].

3) a) When light strikes a pixel on a CCD, electrons [1 mark] are liberated from the silicon and are stored in a potential well [1 mark]. Once the exposure has been taken, electrons are shunted along the potential wells [1 mark] and emerge in sequence at the output, where they can be measured.

b) CCDs have a quantum efficiency greater than 70% [1 mark] whereas photographic emulsion only has a 4% efficiency. So fewer photons are needed for an image and fainter objects can be detected [1 mark]. The output is in electronic form and can be processed digitally by computers, making it easier for the images to be enhanced [1 mark].

Page 57 — Non-Optical Telescopes

1) The collecting power of the telescope is proportial to the area of the objective dish or mirror [1 mark]. Radio telescopes tend to have larger dishes than UV telescopes, so radio telescopes tend to have greater collecting powers [1 mark]. Resolving power depends on the wavelength of the radiation and the diameter of the dish [1 mark]. Since UV radiation has a much, much smaller wavelength than radio, UV telescopes have a greater resolving power. [1 mark]

2) a) The telescope emits infrared radiation, which masks the infrared it is trying to detect [1 mark]. The colder the telescope, the less infrared it emits [1 mark].

b) They are set up at high altitude in dry places [1 mark].

3) a) On high altitude aeroplanes / weather balloons [1 mark], to get above the level of the atmosphere that absorbs the radiation [1 mark].

b) A UV telescope uses a single parabolic mirror, whereas an X-ray telescope uses a series of nested 'grazing' mirrors [1 mark]. This is because UV reflects in the same way as visible light [1 mark] but X-rays can only be reflected at very shallow angles / would be absorbed by a parabolic mirror [1 mark].

4) a) power ∝ diameter^2 [1 mark].

b) $\frac{\text{power of Arecibo}}{\text{power of Lovell}} = \frac{300^2}{76^2}$ [1 mark]. Ratio = 15.6:1 [1 mark].

Page 59 — Distances and Magnitude

1) The absolute magnitude is the apparent magnitude [1 mark] that the object would have if it were 10 parsecs [1 mark] away from Earth.

2) Distance to Sun in parsecs $= 1/(2 \times 10^5) = 5 \times 10^{-6}$ pc [1 mark].

$\mathbf{m} - \mathbf{M} = 5 \lg(\mathbf{d}/10)$ [1 mark] $\Rightarrow -27 - \mathbf{M} = 5 \lg(5 \times 10^{-6}/10)$ [1 mark]

$\Rightarrow -27 - \mathbf{M} = 5 \lg(5 \times 10^{-7}) \Rightarrow -27 - \mathbf{M} = -31.5 \Rightarrow \mathbf{M} = 4.5$ [1 mark].

3) a) Sirius is the brighter of the two [1 mark].

b) $\mathbf{m} - \mathbf{M} = 5 \lg(\mathbf{d}/10)$ [1 mark] $\Rightarrow -0.72 - (-5.5) = 5 \lg(\mathbf{d}/10)$ [1 mark]

$\Rightarrow 4.78 = 5 \lg(\mathbf{d}/10) \Rightarrow \lg(\mathbf{d}/10) = 0.956 \Rightarrow \mathbf{d}/10 = 10^{0.956}$

So $\mathbf{d} = 90$ pc [1 mark].

Page 61 — Stars as Black Bodies

1) a) According to Wein's displacement law $\lambda_{max} \times \mathbf{T} = 0.0029$, so for this star $\lambda_{max} = 0.0029 \div 4000 = 7.25 \times 10^{-7}$ m [1 mark]. Curve Y peaks at around 0.7 μm ($= 7 \times 10^{-7}$ m), so could represent the star [1 mark].

b) $\mathbf{L} = \sigma\mathbf{A}\mathbf{T}^4$, so $3.9 \times 10^{26} = 5.67 \times 10^{-8} \times \mathbf{A} \times 4000^4$ [1 mark], which gives $\mathbf{A} = 2.7 \times 10^{19}\ \text{m}^2$ [1 mark].

2) $\lambda_{max} \times \mathbf{T} = 0.0029$ [1 mark]. So $\mathbf{T} = 0.0029/(436 \times 10^{-9}) \approx 6650$ K [1 mark].

$\mathbf{L} = \sigma\mathbf{A}\mathbf{T}^4$ [1 mark]. So $2.3 \times 10^{27} = 5.67 \times 10^{-8} \times \mathbf{A} \times 6650^4$, which gives $\mathbf{A} = 2.1 \times 10^{19}\ \text{m}^2$ [1 mark].

Page 63 — Spectral Classes and the H-R Diagram

1) a) To get strong Balmer lines, the majority of the electrons need to be at the $\mathbf{n} = 2$ level [1 mark]. At low temperatures, few electrons have enough energy to be at the $\mathbf{n} = 2$ level [1 mark]. At very high temperatures, most electrons are at $\mathbf{n} = 3$ or above, leading to weak Balmer lines [1 mark].

b) Spectral classes A [1 mark] and B [1 mark]

c) Class F stars are white [1 mark], have a temperature of 6000 – 7500 K [1 mark] and show prominent absorption lines from metal ions [1 mark].

2) Molecules are only present in the lowest temperature stars [1 mark]. At higher temperatures molecules are broken up into individual atoms [1 mark].

3)

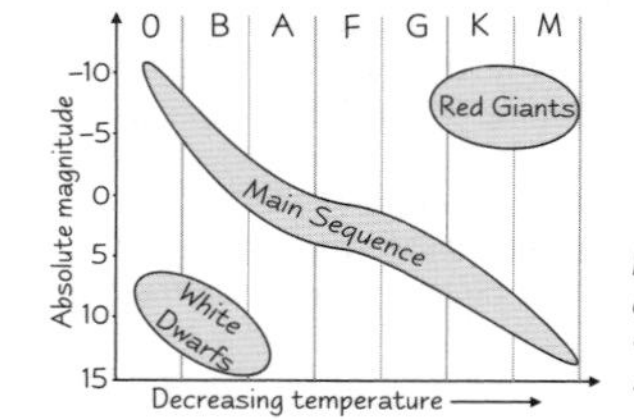

[5 marks maximum, 1 mark each for correctly labelled axes, 1 mark each for 'Main Sequence', 'White Dwarfs' and 'Red Giants'.]

Answers

Page 65 — Stellar Evolution

1) High mass stars spend less time on the main sequence than low mass stars [1 mark]. As red giants, low mass stars only fuse hydrogen and/or helium but the highest mass stars fuse nuclei up to iron [1 mark]. Lower mass stars eject their atmospheres to become white dwarfs [1 mark], but high mass stars explode in supernovae [1 mark] to leave neutron stars [1 mark] or black holes [1 mark].

2) a) The Schwarzschild radius is the distance [1 mark] from the centre of a black hole to where the escape velocity is the speed of light [1 mark].

b) $R_s = \frac{2GM}{c^2} = \frac{2 \times 6.67 \times 10^{-11} \times 6 \times 10^{30}}{(3 \times 10^8)^2}$ [1 mark] ≈ 8.9 km [1 mark]

Page 67 — The Doppler Effect and Redshift

1) a) Object A is moving towards us [1 mark].

b) Object B is part of a binary star system (or is being orbited by a planet) [1 mark] with a period of two weeks [1 mark].

c) Find the velocity of object C using $\frac{\Delta\lambda}{\lambda} = \frac{v}{c}$, so $v = c\frac{\Delta\lambda}{\lambda}$. [1 mark]

$$v = 3.0 \times 10^8 \times \frac{(667.83 \times 10^{-9} - 656.28 \times 10^{-9})}{656.28 \times 10^{-9}} = 5.28 \times 10^6 \, ms^{-1}$$

So object C is moving away from us [1 mark] at 5.28×10^6 ms^{-1} [1 mark].

2) a) Their spectrum shows an enormous redshift [1 mark].

b) Intensity is proportional to 1/distance^2 [1 mark]. So, e.g., if a quasar is 500 000 times further away than, but just as bright as, a star in the Milky Way it must be 500 000^2 times brighter than the star [1 mark].

c) A supermassive black hole [1 mark] surrounded by a doughnut-shaped mass of whirling gas [1 mark].

Page 69 — The Big Bang Model of the Universe

1) a) $v = H_0 d$ [1 mark] where v is recessional velocity (in kms^{-1}), d is distance (in Mpc) and H_0 is Hubble's constant in (kms^{-1}Mpc^{-1}). [1 mark]

b) Hubble's law suggests that the Universe originated with the Big Bang [1 mark] and has been expanding ever since. [1 mark]

c)i) $H_0 = v \div d$ = 50 kms^{-1} ÷ 1 Mpc^{-1}.
50 kms^{-1} = 50 × 10^3 ms^{-1} and 1 Mpc^{-1} = 3.09 × 10^{22} m
So, H_0 = 50 × 10^3 ms^{-1} ÷ 3.09 × 10^{22} m = 1.62 × 10^{-18} s^{-1}
[1 mark for the correct value, 1 mark for the correct unit]

ii) $t = 1/H_0$ [1 mark] = 1/1.62 × 10^{-18} = 6.18 × 10^{17} s ≈ 2 × 10^{10} years [1 mark]
The observable Universe has a radius of ~20 billion light years. [1 mark]

2) a) $z \approx v/c$ [1 mark] so v ≈ 0.37 × 3.0 × 10^8 ≈ 1.1 × 10^8 ms^{-1} [1 mark]

b) $d = v/H_0$ ≈ 1.1 × 10^8 / 2.4 × 10^{-18} = 4.6 × 10^{25} m [1 mark]
= 4.6 × 10^{25} / 9.5 × 10^{15} ly = 4.9 billion ly [1 mark]

c) $z = v/c$ is only valid if $v << c$ — it isn't in this case [1 mark].

3) The cosmic background radiation is microwave radiation [1 mark] showing a perfect black body spectrum [1 mark] of a temperature of about 3 K [1 mark]. It is very nearly isotropic and homogeneous [1 mark].
It suggests that the ancient Universe was very hot, producing lots of electromagnetic radiation [1 mark] and that its expansion has stretched the radiation into the microwave region [1 mark]. [1 mark for quality of written communication.]

Unit 5: Option B — Medical Physics

Page 71 — Physics of the Eye

1) a) The distance will be the focal length of the lens [1 mark]
$v = f$ = 1/power = 1/60 = 0.017 m [1 mark]

b) $1/u + 1/v = 1/f$, u = 0.3 m, v = 1/60 m [1 mark].
$1/f$ = 63.3 D [1 mark]. So the extra power needed = 3.3 D [1 mark].

2) a) Light enters the eye through the cornea, which focuses the light [1 mark]. It then passes through the aqueous humour, then through the iris, which controls the amount of light entering the eye [1 mark], to reach the lens. The lens acts as a fine focus [1 mark] so that light travels through the vitreous humour and is focused on the retina [1 mark].

b) Spatial resolution is greatest at the yellow spot on the retina where the photoreceptors are most densely packed [1 mark]. Away from the yellow spot, spatial resolution decreases with receptor density, and because of an increase in the number of receptors per nerve cell [1 mark].

Page 73 — Defects of Vision

1) Focal length of diverging lens needs to be –4 m [1 mark].
Power = $1/f$ = –0.25 D [1 mark for value, 1 mark for negative sign]

2) Lens equation $1/u + 1/v = 1/f$ [1 mark]. When u = 0.25 m, v = –2 m
⇒ $1/f$ = 1/0.25 – 1/2 = 3.5. Power = +3.5 D [1 mark for value, 1 mark for sign]

3) a) Cylindrical lenses [1 mark].

b)

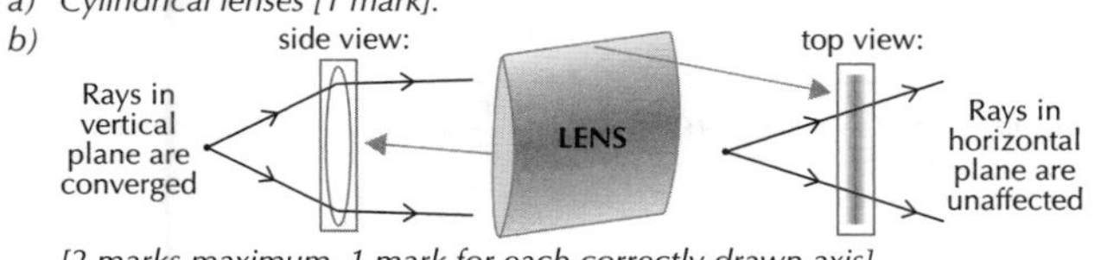

[2 marks maximum, 1 mark for each correctly drawn axis]

Page 75 — Physics of the Ear

1) a) The pinna concentrates the sound energy entering the ear into the auditory canal, increasing its intensity [1 mark].

b) Sound energy entering the ear causes the tympanic membrane (eardrum) to vibrate [1 mark]. The vibrations are transmitted through the middle ear by the malleus, incus and stapes in turn [1 mark]. The stapes is connected to the oval window, so causes it to vibrate [1 mark].

c) The amplitude of a sound is proportional to the square root of its intensity, and the intensity is inversely proportional to the area [1 mark]. This means that the amplitude is inversely proportional to the square root of the area [1 mark], so if the area is decreased by a factor of 14, the amplitude is increased by a factor of $\sqrt{14} \approx 3.74$ [1 mark].

d) Pressure waves in the cochlea cause the basilar membrane to vibrate [1 mark], which causes hair cells on the membrane to trigger electrical impulses [1 mark].

e) Different regions of the basilar membrane have different natural frequencies [1 mark]. When the frequency of a sound wave matches the natural frequency of a part of the membrane, that part resonates, causing the hair cells in that area to trigger impulses, so different frequencies trigger different nerve cells [1 mark].

Page 77 — Intensity and Loudness

1) a) 1 × 10^{-12} Wm^{-2} [1 mark]

b) $IL = 10\log\left(\frac{I}{I_0}\right)$ [1 mark] $= 10\log\left(\frac{0.94}{1 \times 10^{-12}}\right) = 119.7$ dB [1 mark]

c) The ear is most sensitive at about 3000 Hz, so the siren will sound as loud as possible [1 mark].

2) The patient has suffered hearing loss at all frequencies, but the loss is worst at high frequencies [1 mark]. If the patient's hearing had been damaged by excessive noise, you would expect to see a peak at a particular frequency [1 mark]. This isn't present, so the patient's hearing loss is more likely to be age-related [1 mark].

Page 79 — Physics of the Heart

1) a) Dead skin cells and hairs are removed [1 mark]. Conductive gel is used to give a good electrical contact [1 mark].

b) The P wave corresponds to depolarisation and contraction of the atria [1 mark]. The QRS wave corresponds to depolarisation and contraction of the ventricles [1 mark] and repolarisation and relaxation of the atria [1 mark].

2) A membrane is initially polarised so that the outside is positively charged and the inside is negatively charged [1 mark]. When the membrane is stimulated, it becomes permeable to sodium ions [1 mark]. The ions move through the membrane and into the cell, depolarising the system [1 mark] and then polarising it the other way. In repolarisation, the membrane becomes impermeable to sodium, but very permeable to potassium ions [1 mark].
The potassium ions move through the membrane to reverse the polarisation, and the Na-K pump moves sodium ions out of the cell to restore equilibrium [1 mark]. [1 mark for quality of written communication]

3) The sinoatrial node produces ~70 electrical pulses a minute [1 mark]. These make the atria contract [1 mark]. They then pass to the atrioventricular node, which delays the pulses [1 mark] then passes them to the ventricles to make them contract shortly after [1 mark].

Page 81 — X-ray Production

1)

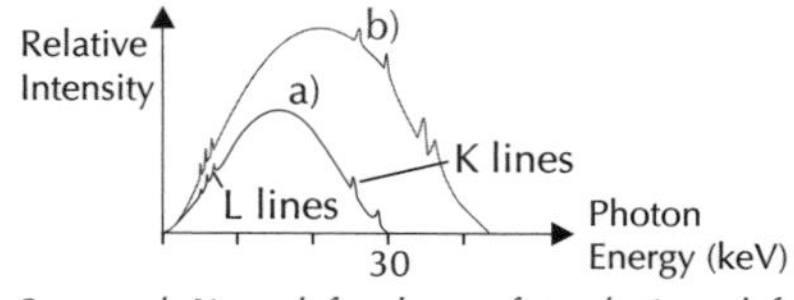

a) See graph [1 mark for shape of graph, 1 mark for 30 keV maximum energy and 1 mark for correct labelling of line spectrum]

b) See graph [1 mark for higher intensity and higher maximum energy and 1 mark for a few extra lines in the line spectrum]

c) Find the energy of each electron using $E = QV$:
E = 1.6 × 10^{-19} × 30 × 10^3 = 4.8 × 10^{-15} J [1 mark]
Kinetic energy = ½mv^2 = 4.8 × 10^{-15} [1 mark]
So, v^2 = (2 × 4.8 × 10^{-15}) ÷ (9.11 × 10^{-31}) = 1.054 × 10^{16} [1 mark]
v = 1.03 × 10^8 ms^{-1} [1 mark]

Page 83 — X-ray and MRI Imaging

1)a) Half-value thickness is the thickness of material required to reduce the intensity of an X-ray beam to half its original value [1 mark].

b) $\mu = \frac{\ln 2}{x_{\frac{1}{2}}} = \frac{\ln 2}{3} = 0.23$ mm^{-1} [1 mark], $I = I_0 e^{-\mu x} \Rightarrow \frac{I}{I_0} = e^{-\mu x}$ [1 mark].
So, 0.01 = $e^{-0.23x}$ ⇒ ln (0.01) = –0.23x [1 mark], x = 20mm [1 mark].

Answers

2) The patient lies in the centre of a large magnet, which produces a magnetic field [1 mark]. The magnetic field aligns hydrogen protons in the patient's body [1 mark]. Radio frequency coils are used to transmit radio waves, which cause the aligned protons to absorb energy [1 mark]. When the radio waves stop the protons emit the stored energy as radio waves, which are recorded by the scanner [1 mark]. A computer analyses the received radio waves to produce an image of the patient's body [1 mark]. [1 mark for quality of written communication]

3) Advantages, e.g. there are no known side effects / doesn't use ionising radiation / an image can be made for any slice in any orientation of the body / images of soft tissue are higher quality than using other techniques (CT, X-ray) / contrast can be weighted to investigate different situations.
Disadvantages, e.g. other techniques give better quality images of bony structures / people can suffer claustrophobia inside the scanner / MRI cannot be used on people with pacemakers/surgical implants/other metal in their bodies / MRI scanners are very expensive.
[1 mark for each point explained, to a total of 6 marks]

Page 85 — Ultrasound Imaging

1) a) $\alpha = \left(\frac{Z_{tissue} - Z_{air}}{Z_{tissue} + Z_{air}}\right)^2 = \left(\frac{1630 - 0.430}{1630 + 0.430}\right)^2$ [1 mark], $\alpha = 0.999$ [1 mark]

b) From part a), 0.1% enters the body when no gel is used [1 mark].

$\alpha = \left(\frac{Z_{tissue} - Z_{gel}}{Z_{tissue} + Z_{gel}}\right)^2 = \left(\frac{1630 - 1500}{1630 + 1500}\right)^2 = 0.002$ [1 mark], so 99.8% of the ultrasound is transmitted [1 mark]. Ratio is ~1000 [1 mark].

2) a) $Z = \rho v$ [1 mark], $v = (1.63 \times 10^6)/(1.09 \times 10^3) = 1495\text{ms}^{-1} \approx 1.50\text{ kms}^{-1}$ [1 mark].

b) A pulse from the far side of the head travels an extra $2d$ cm, where d is the diameter of the head [1 mark]. Time taken to travel this distance = 2.4 × 50 = 120 μs [1 mark]. Distance = speed × time, so $2d = 1500 \times 120 \times 10^{-6}$ = 0.18 m [1 mark]. So d = 9 cm [1 mark].

Page 87— Endoscopy

1) a) $\sin \theta_c = n_2/n_1 = 1.30/1.35$ [1 mark], $\theta = 74.4°$ [1 mark].

b) When the angle of incidence is greater than or equal to the critical angle, the beam of light will undergo total internal reflection [1 mark]. If the angle of incidence falls below the critical angle, then some light will be lost [1 mark].

2) A coherent fibre-optic bundle consists of a large number of very thin fibres [1 mark], arranged in the same way at either end of the bundle [1 mark]. Lots of thin fibres are used to increase the resolution of the image [1 mark]. The relative positions of the fibres have to remain constant or the image would be jumbled up [1 mark].

Unit 5: Option D — Turning Points in Physics

Page 89 — Charge/Mass Ratio of the Electron

1) a) 1000 eV [1 mark]

b) 1000 eV × 1.6×10^{-19} J/eV = 1.6×10^{-16} J [1 mark]

c) Kinetic energy = $\frac{1}{2}mv^2 = 1.6 \times 10^{-16}$ J [1 mark]
$v^2 = (2 \times 1.6 \times 10^{-16}) \div (9.1 \times 10^{-31}) = 3.5 \times 10^{14} \Rightarrow v = 1.9 \times 10^7\text{ ms}^{-1}$ [1 mark]
Divide by 3.0×10^8: 6.3% of the speed of light [1 mark]

2) [Your answer will depend on which experiment you describe, e.g.]
Electrons are accelerated using an electron gun [1 mark]. A magnetic field [1 mark] exerts a centripetal force [1 mark] on the electrons, making them trace a circular path. By measuring the radius of this path and equating the magnetic and centripetal forces [1 mark] you can calculate e/m. [1 mark for quality of written communication]

Page 91 — Millikan's Oil-Drop Experiment

1) a) The forces acting on the drop are its weight, acting downwards [1 mark] and the equally sized force due to the electric field, acting upwards [1 mark].

b) Weight = electric force, so $mg = \frac{QV}{d}$, and $Q = \frac{mgd}{V}$ [1 mark].

$Q = \frac{1.63\times10^{-14}\times9.81\times3.00\times10^{-2}}{5000} = 9.59\times10^{-19}C$ [1 mark]

Divide by e: $9.59 \times 10^{-19} \div 1.6 \times 10^{-19} = 6 \Rightarrow Q = 6e$ [1 mark]

c) The forces on the oil drop as it falls are its weight and the viscous force from the air [1 mark]. As the oil drop accelerates, the viscous force increases until it equals the oil drop's weight [1 mark]. At this point, there is no resultant force on the oil drop, so it stops accelerating, but continues to fall at terminal velocity [1 mark].

d) At terminal velocity, $mg = 6\pi\eta rv$. Rearranging, $v = \frac{mg}{6\pi\eta r}$ [1 mark]

Find the radius of the oil drop, using mass = volume × density:

$m = \frac{4}{3}\pi r^3 \rho$. So, $r^3 = \frac{3m}{4\pi\rho} = \frac{3\times1.63\times10^{-14}}{4\times\pi\times880} = 4.42\times10^{-18}$

and $r = 1.64 \times 10^{-6}$ m [1 mark].

So, $v = \frac{1.63\times10^{-14}\times9.81}{6\times\pi\times1.84\times10^{-5}\times1.64\times10^{-6}} = 2.81\times10^{-4}\text{ms}^{-1}$ [1 mark]

Page 93 — Light — Newton vs Huygens

1) Light consists of particles [1 mark]. The theory was based on Newton's laws of motion with the straight-line motion of light as evidence [1 mark].

2) Most scientists in the 18th century supported Newton's corpuscular theory [1 mark]. He said that light was made up of particles that obey his laws of motion [1 mark]. In Huygens' wave theory, light is a wave [1 mark]. This is supported by the diffraction and interference seen in Young's double-slit experiment [1 mark]. In the second half of the 19th century, Maxwell described light as an electromagnetic wave [1 mark] consisting of oscillating electric and magnetic fields. [1 mark for quality of written communication]

Page 95 — The Photoelectric Effect

1) Electrons in the metal absorb energy from the UV light and leave the surface [1 mark], causing the plate to become positively charged [1 mark].

2) a) $\phi = hf - (\frac{1}{2}mv^2)_{max} = hc/\lambda - (\frac{1}{2}mv^2)_{max}$ [1 mark]
$= ((6.63 \times 10^{-34} \times 3.00 \times 10^8)/(0.5 \times 10^{-6})) - 2.0 \times 10^{-19}$ J [1 mark]
$= 1.98 \times 10^{-19}$ J [1 mark]

b) Electrons will only be emitted if the energy they gain from photons in the beam of light is greater than the work function energy [1 mark].
The energy supplied by the beam of light is:
$E = hc/\lambda = (6.63 \times 10^{-34} \times 3.00 \times 10^8)/(1.5 \times 10^{-6}) = 1.33 \times 10^{-19}$ J [1 mark], which is less than the work function calculated in (a), so will not supply enough energy for electrons to be released [1 mark].

3) $\phi = 2.2\text{ eV} = 2.2 \times 1.6 \times 10^{-19}\text{ J} = 3.52 \times 10^{-19}$ J
$(\frac{1}{2}mv^2)_{max} = hf - \phi = (hc/\lambda) - \phi$ [1 mark]
$= [6.63 \times 10^{-34} \times (3.00 \times 10^8/350 \times 10^{-9})] - 3.52 \times 10^{-19}$
$= 2.16 \times 10^{-19}$ J [1 mark]
Stopping potential, $V_s = (\frac{1}{2}mv^2)_{max}/e$ [1 mark]
$\Rightarrow V_s = 2.16 \times 10^{-19}/1.6 \times 10^{-19}\text{ C} = 1.35$ V [1 mark]

Page 97 — Wave-Particle Duality

1) a) i) Velocity is given by $\frac{1}{2}mv^2 = eV$ [1 mark]
$\Rightarrow v^2 = 2eV/m \Rightarrow v = 1.3 \times 10^7\text{ ms}^{-1}$ [1 mark]
ii) de Broglie $\lambda = h/mv$ [1 mark] $\Rightarrow \lambda = 5.5 \times 10^{-11}$ m [1 mark]

b) This is in the X-ray region of the EM spectrum [1 mark].

2) a) A stream of electrons is accelerated towards the sample using an electron gun [1 mark]. The beam of electrons is focused onto the sample using a magnetic field [1 mark]. The parts of the beam that pass through the sample are projected onto a screen to form an image of the sample [1 mark].

b) To resolve detail around the size of an atom, the electron wavelength needs to be around 0.1 nm [1 mark]. The relationship between anode voltage and electron wavelength is given by $\lambda = \frac{h}{\sqrt{2meV}}$,

which rearranges to give $V = \frac{h^2}{2me\lambda^2}$ [1 mark].

Substituting $m = 9.1 \times 10^{-31}$ kg, $e = 1.6 \times 10^{-19}$ C, $l = 0.1 \times 10^{-9}$ m gives:

$V = \frac{(6.63\times10^{-34})^2}{2\times9.1\times10^{-31}\times1.6\times10^{-19}\times(0.1\times10^{-9})^2} = 151\text{ V}$,

showing that the minimum anode voltage ≈ 150 V [1 mark].

Page 99 — The Speed of Light and Relativity

1) a) The interference pattern would move/be shifted [1 mark].

b) The speed of light has the same value for all observers [1 mark].
It is impossible to detect absolute motion [1 mark].

2) a) An inertial reference frame is one in which Newton's 1st law is obeyed [1 mark], e.g. a train carriage moving at constant speed along a straight track (or any other relevant example) [1 mark].

b) The speed of light is unaffected by the motion of the observer [1 mark] or the motion of the light source [1 mark].

Page 101 — Special Relativity

1) $t = \frac{t_0}{\sqrt{1 - \frac{v^2}{c^2}}}$ [1 mark] and $t_0 = 20 \times 10^{-9}$ s [1 mark]

$t = \frac{20\times10^{-9}}{\sqrt{1 - \frac{(0.995c)^2}{c^2}}} = 200$ ns or 2×10^{-7} s [1 mark]

2) Your description must include:
A diagram or statement showing relative motion [1 mark].
An event of a specified duration in one reference frame [1 mark].
Measurement of the time interval by a moving observer [1 mark].
Time interval for "external" observer greater than time interval for the "stationary" observer or equivalent [1 mark].

3) a) $m = m_0 \div \sqrt{1 - \frac{v^2}{c^2}} = 1.67\times10^{-27} \div \sqrt{1 - \frac{2.8^2}{3.0^2}} = 4.65\times10^{-27}$ kg [1 mark]

b) $E = mc^2 = 4.65 \times 10^{-27} \times (3 \times 10^8)^2 = 4.2 \times 10^{-10}$ J [1 mark]

Index

A

absolute
- magnitude of a star 59, 63
- motion 98
- temperature 46, 50
- zero 46

absorption spectra 62, 66
acceleration 5
- centripetal 7
- due to gravity 15
- simple harmonic motion 8, 9

acoustic impedance 84, 85
action potential 78
age of the Universe 68
Airy disc 53
alpha particles 32, 36, 40, 41
alternators 30, 31
amplitude 12
- ultrasound scan (A-Scan) 85
- simple harmonic motion 9-11
- sound 74, 75

angular magnification 52
angular speed 6, 7
anode 88
apparent magnitude 58, 59
aqueous humour 70
astigmatism 72
astrophysics unit (AU) 59
atmosphere transparency 56
atomic
- mass units 42
- nuclei 34, 35, 40, 42
- number 40
- structure 33

atria 78
attenuation coefficient 82
auditory nerve 74, 75
Avogadro's constant 47

B

background radiation 37, 38
Balmer series 62, 67
barium meal 82
beta radiation 36, 40, 41
binary stars 66
binding energy 42, 43, 45
black body radiation 60, 69
black holes 64, 65, 67
blueshift 66
Boltzmann constant 50
Boyle's law 46
bremsstrahlung 80
brightness
- of a star 58
- ultrasound scan (B-Scan) 85

C

capacitors 22-25
- time constant 25

Captain Skip 42
Cassegrain arrangement 53
cathode
- ray oscilloscopes 85
- rays 88

centripetal forces 7, 16, 89
charge-coupled devices (CCDs) 54, 56
charge/mass ratio of the electron 89
Charles' law 46
chromatic aberration 53
circular motion 6, 7, 16, 17, 27, 89
cochlea 74, 75
coherent
- light sources 93
- optical fibres 86

collecting power of a telescope 57
collisions 4, 48-50
colour perception 71
computed tomography (CT) scans 82
conclusions 3
cone cells 70, 71
conservation
- of energy 29, 41
- of momentum 4, 41, 48

control rods 44
controlled experiments 3
coolant 44
cornea 70
corpuscular theory of light 92
cosmic microwave background radiation (CMBR) 69
cosmic radiation 37
cosmological principle 68, 69
cosmological redshift 66
Coulomb's law 18
critical
- angle 86
- damping 13
- mass 44

cyclotron 6

D

damping 12, 13
dark energy 68
de Broglie
- equation 96
- wavelength 96, 97

decay constant 38
decibels 76
dielectric 24
diffraction 33, 93, 96, 97
dioptres 52
displacement (simple harmonic motion) 8, 9
Doppler effect 66, 67
double-slit experiment 93
driving frequency 12
dynamos 31

E

ears 74-76
eddy currents 30
efficiency (of a transformer) 30
Einstein, Albert 94, 99
elastic collisions 4, 49
electric
- charge 18
- fields 18-20
- potential 20

electrocardiogram (ECG) 79
electromagnetic induction 28-31
electromagnetic waves 93
electromotive force (e.m.f.) 28, 29, 31
electron
- beams 88
- capture 41
- degeneracy pressure 64
- diffraction 33, 96
- gun 88, 89
- microscopes 97

electronvolt (eV) 88
electrons 80, 82, 88, 89, 91, 94-97
- charge on 91
- charge/mass ratio 88, 89
- discovery of 88

electrostatic repulsion in nuclei 40, 45
emission spectral lines 62, 67
endoscopy 86, 87
energy
- conservation of 29, 41
- kinetic 4, 8, 17, 50, 51, 88, 94, 95
- levels in atomic hydrogen 62
- potential 8, 16, 17, 20

equal loudness curves 77
event horizon 65
evidence 2, 3
explosions 4
extra-solar planets 66
eyes 70-72

F

far point 70, 72, 73
Faraday's law 28-30
field lines 14, 19, 21, 26
field strength 15, 18-20
fields
- electric 18-20
- gravitational 14, 15, 21
- magnetic 26-28, 30, 31

fission 42-45
Fleming's left-hand rule 26, 27, 29
fluoroscopy 82
flux
- cutting 28, 31
- density 26
- linkage 28, 31

focal length 52, 53, 70, 72, 73
focal points 55
forced vibrations 12
forces
- electric (Coulomb) 18, 21
- gravitational 14, 15

fovea 70
free vibrations 12
frequency
- Doppler effect 66
- photoelectric effect 94, 95
- simple harmonic motion 6, 9
- sound 76, 84

fusion
- latent heat 51
- nuclear 42, 43, 45, 64

G

galactic nucleus 67
gamma radiation 36, 37, 41
gas laws 46, 47
Geiger-Muller tube 37
generators 31
geosynchronous satellites 17
giant white rabbit 16
gravitational
- fields 14-16, 21
- force 14, 15
- potential 15, 16, 20

grazing telescopes 56

H

half-life 38, 39
hearing loss 77
heart 78, 79
Hertz, Heinrich 93
Hertzsprung-Russell (H-R) diagram 63
homogeneous Universe 68, 69
Hot Big Bang theory 66, 69
Hubble's law 68
Huygens' principle 92
hydrogen fusion 43, 45, 64
hypermetropia (long sight) 72, 73

I

ideal gases 46-50
impulse 5
induced fission 44
induction (electromagnetic) 28-30
inelastic collisions 4
inertial reference frame 99
inner ear 74, 75
intensity
- of radiation 37, 58, 60
- photoelectric effect 94, 95
- reflection coefficient 84
- sound 74-77
- X-ray production 81

interference 92, 93, 96
inverse square law
- 14, 15, 18, 19, 21, 37, 60, 61, 67

ionising properties of radiation 36
iris 70
isotopes 38-40
isotropic Universe 68, 69

K

kelvin 46
kinetic energy 4, 8, 17, 88
- in an ideal gas 50, 51
- photoelectric effect 94, 95

kinetic theory 48, 49

Index

L

latent heat 51
length contraction 101
lens axis 52
lens equation 52, 70, 72, 73
lenses 52, 53, 70, 72, 73
Lenz's law 29
light 92, 93
  photon model 94, 95
  wave-particle duality 96, 97
light-years (ly) 59
linear speed 6, 7
loudness 74, 76, 77
luminosity 58-60, 63, 67

M

magnetic
  fields 26-31, 89
  flux 28, 30
magnetic resonance imaging (MRI) 83
magnification 52, 53, 72
main sequence stars 63, 64
mass 4, 5
  attenuation coefficient 82
  defect 41, 42
  number 34, 40
mass-energy equivalence 41, 42, 101
mass-spring systems 10
Maxwell, James Clerk 93
mean square velocity 48
mechanical energy 8
Michelson-Morley interferometer 98
middle ear 74, 75
Millikan's oil-drop experiment 90, 91
moderators 44
molar gas constant 47
momentum 5
  conservation of 4, 41, 48
  wave-particle duality 96
muon decay (as evidence of time dilation) 100
myopia (short sight) 72, 73

N

national grid 30
natural frequency 12, 75
near point 70, 72
neutron stars 64, 65
neutrons 36, 40, 41, 44, 45
Newton, Isaac 92, 99
Newton's laws 5, 14, 48, 99
nuclear
  decay 40
  density 35
  diameter 33
  fission 42-45
  fusion 42, 43, 45
  radius 34
  reactors 44
nucleons 34, 40-43
nuclei 32-35, 40, 42, 43

O

oil-drop experiment, Millikan's 90, 91
optic nerve 70
optical
  fibres 86
  telescopes 52, 53, 54
orbital period 16, 17
oscillations 8-13
ossicles (malleus, incus, stapes) 74, 75
overdamping 13

P

parallax 58
parsecs 58
particle acclerators 27
peak wavelength 60
peer review 2
penetrating power of radiation 36
period
  of orbit 16
  simple harmonic motion 6, 9-11
permittivity 18, 19, 93
persistence of vision 71
phase difference 12
phons 77
photoelectric effect 54, 93-96
photons 54, 81, 94-96
piezoelectric effect 84
pitch 76
Planck's constant 94, 96
planetary nebulae 64
planets 16
plastic deformation 13
polarisation
  in the heart 78, 79
  of light 93
postulates of special relativity 99
potential 15, 16, 20-22, 24
  difference 22, 24
  energy 8, 16, 17, 20
power of a lens 52, 70, 73
predictions 2
pressure of an ideal gas 46, 48, 50
principle
  axis 52
  focus 52, 73
probability wave 96
proper time 100
proton number 40, 41
protostars 64
pulsars 65
pupil 70

Q

quanta 94
quantised charge 91
quantum efficiency 54
quasars 67

R

radial fields 14, 19, 20
radians 6
radio telescopes 55
radioactive
  decay 36, 38-40
  isotopes 39
radioactivity 36, 38-40
radiocarbon dating 39
radiographers 81
radon gas 37
ray diagrams 52
Rayleigh criterion 53, 55, 57
reactors 44
real images 52, 53, 72
recessional velocity 68
red giants 63, 64
redshift 66-68
reflection 53, 92
refraction 52, 53, 70, 92
refractive index 70, 86
relativistic mass 101
resolving power of a telescope 53, 55, 57
resonance 12, 13, 75
retinas 70, 72
right-hand rule 26
rod cells 70
root mean square (rms) speed 49
Rutherford scattering 32

S

satellites 16, 17
scanning tunnelling microscope (STM) 97
Schwarzchild radius 65
scientific process 2, 3
semicircular canals 74
simple harmonic motion (SHM) 8-11
Snell's law 86
special relativity 99, 100, 101
specific
  charge 89
  heat capacity 51
  latent heat 51
spectra
  absorption 62, 66
  black body 60
  emission 62
  X-ray 80
spectral classes 62, 63
spectroscopic binary stars 66
speed distribution, in a gas 50
speed of light 98-101
spherical aberration 53, 55
spontaneous fission 44
spring constant 10
standard candles 59
stars 58-66
Steady-State Universe 68
Stefan's law 60, 61
stellar evolution 64, 65
stiffness constant 10
Stoke's law 90
stopping potential 95
strong interaction 40, 45
supernovae 59, 64, 65
suspensory ligaments 70

T

tangential velocity 6
telescopes 52-57
temperature 46, 50, 51
teslas 26
theories 2
thermal neutrons 44
thermionic emission 88
Thomson, J. J. 88
threshold frequency 94, 95
threshold of hearing 76
time constant, RC 25
time dilation 100
total internal reflection 86
transformers 30, 31
transmission electron microscope (TEM) 97
transparency (of the atmosphere) 56

U

ultrasound 84, 85
Universe 68, 69
unstable nuclei 36, 40, 41, 44, 45
uranium 40, 43, 44

V

valves 78
velocity 4, 6, 68
  simple harmonic motion 8, 9
ventricles 78, 79
virtual images 52, 72, 73
viscosity 90
vitreous humour 70
voltage, induced 30, 31

W

wave-packets 94
wave-particle duality 96, 97
wavelength 55, 56
  de Broglie 96, 97
  doppler effect 66
wavelets 92
white dwarfs 63, 64
Wien's displacement law 60, 61
work done 20
work function 95

X

X-ray imaging 80-83
X-ray telescopes 56

Y

yellow spot 70
Young, Thomas 92
Young's double slit experiment 93